Thanks for your dedication to Church History!

Craig George

March 2011

On God's Path:
The Unfolding Story of Humanity

S. Craig George

AuthorHouse™
1663 Liberty Drive
Bloomington, IN 47403
www.authorhouse.com
Phone: 1-800-839-8640

© *2010 S. Craig George. All rights reserved.*

No part of this book may be reproduced, stored in a retrieval system, or transmitted by any means without the written permission of the author.

First published by AuthorHouse 6/14/2010

ISBN: 978-1-4520-1983-3 (e)
ISBN: 978-1-4520-1982-6 (sc)

Library of Congress Control Number: 2010905967

Printed in the United States of America
Bloomington, Indiana

This book is printed on acid-free paper.

This book is dedicated to Gus, an enthusiastic nine-year-old with a toothy smile whose dying wish was to be remembered. Although I hardly knew him I am close to some who knew him well. His passing somehow inspired me to pry this manuscript out of a deep rut and get it rolling again on God's path. Gus, we remember.

Thus says the Lord: Stand by the earliest roads, ask the pathways of old
Which is the way to good, and walk it; thus you will find rest for your souls.
                                        Jeremiah 6:16

Our duty, as men and women, is to proceed as if limits to our abilities did not exist. We are collaborators in creation.
                                        Pierre Teilhard de Chardin

The feeling remains that God is on the journey, too.
                                        St. Teresa of Avila

On God's Path:
The Unfolding Story of Humanity

Table of Contents

1. Introduction 1

2. The Genesis of the Cosmos 5
   13.7 Billion to 6 Million B.C.E.

3. The Path of Humanity.. 21
   6 Million to 800 B.C.E.

4. The Axial Age 35
   800 to 200 B.C.E.

5. Jesus... 51
   6 B.C.E. to 27 C.E.

6. Development, Dissemination and Fragmentation ... 59
   200 B.C.E. to 1900 C.E.

7. Assessing the Trajectory of Humanity 71
   1900 C. E. and Beyond

8. Epilog: Playing Our Role in the Story 93

Preface

This book is about the story of humanity. It is a story that has been told with certainty in many widely divergent versions. Despite so many different versions, we would probably all agree that at the root there must be one true story. In our generation the stories seem to turn on two different centers: science and religion. The theme of this book is that the circles of science and religion are not mutually exclusive, but complementary in a much bigger picture. New insights may be available to those who can expand their thinking to connect both circles into a single elegant ellipse which turns on both centers.

From an academic standpoint there is nothing new in this book. Scientific ideas presented are contemporary mainstream and theological ideas are from an orthodox Catholic perspective of mainstream Christianity. Readers who are savvy in one discipline may find ideas in the other discipline novel. Part of the objective of this book is to challenge readers to expand their thinking to embrace the wisdom of both sides and work out for themselves possible convergences.

The author hopes that exposure to some new ideas will be part of the fun of reading this book. It might inspire you to learn more about some ideas that pique your interest. It might cause you to reexamine your faith in a larger context and embrace it in a more meaningful way. It might also challenge you to consider your own place on

God's path. What is your role in advancing the unfolding story of humanity?

Acknowledgements

I am indebted to so many people in helping me navigate through the twists and turns of this project. A few are enumerated here but my sincere thanks extend to many, many more. Gratitude goes to my mentors and friends at Aquinas Institute of Theology, especially Frs. Peddicord, Martin, Bouchard, Steinkerchner, and Srs. Walter, Streeter and Oosdyke. I have shamelessly adapted small golden nuggets of their wise teaching. I will always be grateful to Fr. Pat Eastman who inspired me to make a course correction on my personal path by introducing me to mystics of the past and present. My biggest debt of gratitude, by far, is owed to my wife who has encouraged me to proceed with this work despite large time demands and small promise of any positive outcome. Finally, my respect, admiration and thanks go to the many human voices who quietly and confidently remind us that faith and reason are not natural enemies, but sadly estranged spouses.

Chapter 1

Introduction

Seek, and you will find; ask, and it will be given to you; knock, and the door will be opened for you. For everyone who seeks finds, and everyone who asks receives, and for everyone who knocks, the door will be opened. Matthew 7: 7,8[1]

We humans are drawn to stories. As we grow up our loved ones tell us stories and read us stories from brightly-colored books. As we learn to read our hunger for stories seems to grow. As adults, we are surrounded by stories in the books and magazines we read, the movies and television we watch and the music we listen to. Even the news and sports are unfolding stories.

But how much do we know about the big story, the story of creation and humanity? We probably have a fairly good idea of our current circumstance but how much does that really tell us? If we saw one scene from the middle of a movie or one chapter from the middle of a book, we would learn some things about the story but would have to guess about the beginning and end and the narrative plot that held it all together. If we could see additional scenes or read additional chapters, we could fill in some of the blanks and it would become easier and easier to guess at the blanks in between.

Our understanding of cosmos, humanity and God

is something like this. Like all generations, we have the current chapter open to us. We have some paragraphs from past chapters that have been handed down to us. Each generation is left to fill in the blanks to tell the story from the perspective of their own time. This generation is a little different, though. Through science, we have more elements of the story than humanity has ever had before. We know more about what the beginning of the cosmos was probably like, how the earth was formed, how life developed on earth and how we humans have developed and populated the earth.

Unfortunately, it is difficult to broaden our focus to take in the panoramic view of this story of creation which spans billions of years with one quick look. This short book is an effort to do so. Volumes upon volumes have been written about various phases of the story. Physicists write about the story of the universe. Geologists write about the story of the earth. Biologists write about the story of life. Anthropologists and archeologists write about the story of humanity. Theologians write about the story of the relationship between humans and God. This book seeks to take the big-picture elements of each and weave them together into one big story. But these elements are just part of the story and focus on the present and the past.

A good story generally has a beginning, a middle and an end. Assuming that we are presently situated somewhere in the middle of the story, there will be more story to come. Although this generation has gained a lot more information about the past, there is still the question of the future, the rest of the story. This book avoids speculation about the future in the sense of taking what we know and tacking on a narrative which introduces some possible continuation. It does, however, explore the question of whether the path we are on implies some trajectory for the future.

Now here come the caveats; there are problems with such a big-picture endeavor. Science is mushrooming and it is difficult for anyone to keep up with the hundreds of leading edges of research, much less all of their nuances and implications. In regard to humanity, there is also a wealth of new information about what makes people tick, both psychologically and biologically, which must be reconciled with our own individual observations of the human experience. When it comes to human experience it is sometimes difficult to sort out fact and opinion. There are new archeological and anthropological finds every year which cause us to pause, look back, and reevaluate understandings of our past. These scientific fields are all dynamic. Considering religion, the many ambiguities and conflicts within and among faiths must also be considered. Volumes upon volumes could be written, in fact have been written, on these broad families of topics. Through the widest of panoramic lenses, the lens employed for this book, detail can be distorted or lost entirely. What seems important to one observer may be of little interest to others.

Further, every observer is influenced by the lens of their life experiences that they cannot help but look through. This author looks through lenses of science and theology and presupposes truth in both. That is, legitimate science is presumed to be an important source of human knowledge. Theology, as expressed in various world traditions, is also presumed to contain truth. This author's Catholic Christian lens will easily be recognized as formative in this book. Caveats about distortions and omissions aside, the reader is invited to question and reconcile ideas as presented in this book with their own faith traditions and personal experiences.

For those readers who see truth in science but not faith or truth in faith but not science, be reminded of U.S. Judge Learned Hand's lesson on integrity. "Integrity is a spirit that's not too sure that it's right." In Pope John Paul II's encyclical *Fides et Ratio*[2], he invites the faithful, even encourages us, to participate in an exercise in reconciling truths of faith and science. He reminds us that both science and religion are sources of truth. Since all truths are part of one Truth, all can and will eventually be both reconcilable and reconciled.

We humans like stories, especially stories about ourselves. Come along and listen to a story, your own story. It is a story that spans billions of years and is not yet complete. You may find that it is not only a story that you can read and tell. It is a story that you are part of and a story that you can influence. In this story will we find danger in new understandings of the cosmos? Will we find danger in the beliefs of strangers? Will we unmask a source of danger within ourselves? Or, if we can navigate our way through the dangers, will we discover that we are all precious threads in the tapestry of creation? Will we be challenged to imagine how our lives can make a difference in the story of the cosmos? Let us start by putting preconceived notions aside, opening our minds to new ideas and allowing our hearts to beat with the rhythm of Creation. We are all seekers; let us seek out our story.

Chapter 2

The Genesis of the Cosmos

> *In the beginning, when God created the heavens and the earth, the earth was a formless void, and darkness covered the face of the deep, while a wind from God swept over the waters. Then God said, "Let there be light," and there was light. And God saw that the light was good; and God separated the light from the darkness. God called the light Day and the darkness he called Night. And there was evening and there was morning, the first day.*

This familiar text is, of course, the opening of the Bible, the first five verses of Genesis Chapter 1. According to the Genesis 1 creation account the cosmos including all life on earth was created by God in seven days. This seven-day creation account was accepted as literal truth by countless Jews and Christians for millennia. More recently it has been a source of conscientious debate. In the twentieth century and continuing into the twenty-first a growing mountain of scientific and biblical scholarship defies a literal understanding of the creation account and its implied history of creation of about 6,000 years. In contrast, science is revealing a creation process that has been underway for billions of years and is still, in fact, continuing.

As an introduction to the story of humanity, in this chapter we will focus on the stuff of the universe. We will begin this story with the creation of stars and planets including our own: the sun and the earth. First we will listen to the testimony of science about the genesis of the cosmos. After refreshing our memories about what the biblical accounts say we will ask biblical scholars to help us reconcile the scientific and biblical stories of creation.

The Beginning of the Cosmos

The notion of the "big bang," the explosive starting-point of the universe, is in today's public consciousness. But where did this idea come from and how much confidence should we put in it? Through the last several hundred years we have learned more and more about our cosmos as both telescopes and science have advanced.

A good starting place for a contemporary understanding is with Edwin R. Hubble, the namesake for the space telescope. Hubble, in the first half of the 1900's, observed that the billons of galaxies in the universe beyond our own (and including our own) are moving at great speeds.[3] By "great speeds" I don't mean like a jet plane at 500 miles per hour, but I mean hundreds of thousands of miles per hour. Did you think you were sitting still reading this book? Far from it. We spin on the surface of the earth at about 600 miles per hour. The earth rotates around the sun at about 60,000 miles an hour. Our solar system spins in our spiral galaxy in excess of 100,000 miles per hour and our galaxy speeds away from the center of the universe at over 200,000 miles per hour.

Among Hubble and those who followed him, the movement of a large number of galaxies was studied. Scientists mapped their position, their speed and their current direction. Then, by projecting this movement backwards in time and space, the scientists could determine their historical path. It turns out that by plotting the paths of a large number of galaxies, it appeared that they all came from the same place at the same time. The time is about 13.7 billion years ago. And the place is roughly the center of the universe as we understand it.

Although this approach of projecting backwards might sound too complicated to put a lot of confidence in, the concept it is quite simple. Imagine attending a night baseball game in which the stadium is having some major lighting problems. Let's say that the lights are such that from your seat you can see the outfield perfectly well, but the infield seems mostly dark and you can't see the bases or home plate. By watching a large number of both fly balls and grounders hit into the outfield, could you figure out about where home plate is located? Sure you could. The physicists just work with a lot bigger ball park. In addition to this projecting back exercise, there are other scientific findings which support the generally accepted time and place of the big bang. There are actually measurable radiation traces left over from the bang itself.

To get a sense of perspective of what this means we must take a step back, an almost unimaginably large step back. Science tells us that this beginning of the universe, the big bang, occurred somewhere in the neighborhood of 13.7 billion years ago. But how can we get our mind around a number so large? Let's start small. Go to your kitchen cabinet and get out your box of toothpicks and find a ruler or measuring tape. Lay some of the toothpicks out side-by-side, like a little log path, and count the number of toothpicks in one inch. I count about a dozen. For purposes of this thought-picture, we'll let each toothpick represent one year. A full human lifetime would be represented by seven or so inches of toothpick path.

We probably don't have enough toothpicks in our kitchen to go back anywhere near the big bang, but we can lay out 13.7 billion toothpicks in our minds. Let's start at a particular place so we have the same picture in mind. This is, after all, a religious book, and written in the United States, so let's choose as our beginning point, representative of today, the center of the alter in St. Patrick's Cathedral in New York City. From this point we will lay out our toothpick path heading generally west across the U.S and beyond.

How far west do we need to go? You might need the help of a junior high school student here, but the math is straightforward. Divide 13.7 billion years by 12 toothpicks per inch, then by 12 inches per foot, then by 5280 feet per mile and we find out our toothpick path must be about 18,000 miles long. How far is that? The circumference of the earth is about 25,000 miles, so our 13.7 billion-toothpick path will wrap almost three-quarters of the way around the earth. With a string and a globe and your junior high student you can easily do this yourself. Starting in New York going west at about the same latitude our string crosses the U.S., Pacific Ocean, China, Turkey, the Mediterranean, and ends in Spain. That's the story for the age of the Cosmos as we understand it now. The toothpick path starts in New York City and extends west all the way around the world ending in Spain.

The idea of the big bang as the beginning of creation might raise a lot of questions for you. You might wonder, was there anything before the big bang? Scientists don't know, but are working on trying to figure that out now. It may be illogical but it's not impossible. You might also wonder, will the expansion of the universe always continue? Scientists aren't sure about that, either: it might continue to expand forever, it might stop expanding at some time, or it might actually start contracting someday.[4] But not to worry, it appears that the expansion is likely to continue for billions upon billions of years, so we have lots of time to figure that out. You might wonder, are there other universes? While we have no way to know that right now, some scientist point out that their equations of physics don't preclude the existence of other universes.

The Story of the Earth Begins

Back to our own universe, what do scientists think transpired after the big bang; how did we get from there to here, from then to now? In the huge explosion of the big bang, time, space and the building blocks of matter came into existence. As cooling began patchy clouds of the lightest atoms probably started to appear. As gravitational attraction started its work, balls of rotating gas resulted in the fusion of atoms. Chain reactions resulted in the formation of stars. Swarms of stars influenced each other through gravitational attraction and galaxies were formed. Stars developed life cycles, burning brightly at first, progressively less brightly, collapsing into themselves and exploding into new stars and matter. Loose matter around stars sometimes came together as the small fragments of material rotated around the closest star. This accretion of matter formed planets like our own planet, earth.[5]

Let's use the same mental image of a toothpick path to think about our home planet. How does our home fit into the big picture of the cosmos; how old is earth? Science tells us that the earth is about 4.6 billion years old, about one-third as old as the cosmos. Let's go back to the toothpicks to get some perspective. Referring back to our toothpick path beginning in St. Patrick's in New York, the segment of the path that corresponds to the age of our planet extends across the continental U.S. and across the Pacific Ocean ending just short of Japan, roughly one-third of the total path.

After our planet formed, what happened next? The earth was at first just a ball of molten rock. After a time of cooling, a crust developed. This crust is what we know as the earth's surface. There is still molten rock miles beneath our feet acting like a boiling pot, an oven which causes deep convection currents to move the crust. The crust is made up of a number of segments, referred to by scientists as plates, and the field of study of this mechanism is called plate tectonics. The continents are made up of subcontinental plates, still shifting, although the movement is slow, often only inches of movement per year. Except for earthquakes which relieve pressure between continental plates slowly grinding against each other, we would not realize it.

Volcanoes also demonstrate the release of the molten rock which exists below us. It appears that some periods in the earth's history had more volcano activity and some had less. Periods of high volcanic activity are often followed by significant global cooling as the clouds of chemicals emitted by the volcanoes significantly block sunlight.

Back to the story of the development of our planet,

after the earth cooled sufficiently, steam in the atmosphere condensed into water and oceans were formed. It is believed that about a billion years after the earth was formed, water had condensed and simple chemical compounds which were the precursors to simple cells started to develop on the ocean floor. After another 3 billion years, just 600 million years ago, the first simple creatures which were pre-insects and pre-crustaceans showed up on the earthly stage. From this point in our planets history on, we have the fossil record and a number of different dating systems to employ to determine the order of development of life. The branch of science which studies these fossil records is called paleontology. According to paleontologists, fish and pre-amphibians swam onto the stage of life on earth about 500 million years ago, about the same time as land plants showed up. Reptiles followed about 300 million years ago and the great reptiles, the dinosaurs, began their age about 220 million years ago. The earliest mammals showed up about 200 million years ago followed by early birds about 150 million years ago.[6]

I have generally avoided the word *evolution* purposely since it seems to be emotionally charged with the vitriol of human argument. Without using this word which can carry a connotation of what caused things to happen as they did, we can still discuss the generally accepted scientific view of the unfolding story of life and ideas about the timing of what happened. One piece of the argument that is worthy of note is the technology of determining the age of things from the past. In tracing the slow progression of different forms of life on our planet, it is fair to say that arguments which attempt to discount one or another technique of aging rocks or fossils have largely been debunked. There are now numerous aging techniques and they tend to reinforce each other. This, of course, adds weight to the generally accepted view of a million upon million year process of development of life on earth as described above.

Before moving along, let's revisit our toothpick path and locate the development of life on earth on our mental line of toothpicks. If you are getting bored with the toothpicks or the many stages of development, fear not. An important point is on the horizon. Recall the path representing the time since the big bang stretched from New York going west all the way around the world to Spain. Heading back along the path, formation of earth occurs about two-thirds of the way back, after we pass Japan. The place on the path relating to the first simple creatures on earth, 600 million toothpicks or years, brings us back to a point only 789 miles west of New York City, a point just east of St. Louis.

Next we'll trace this same route back to New York City noting some important mile markers on our toothpick path along the way. Fish and pre-amphibians showed up about 500 million years ago, a little farther west than Indianapolis, Indiana. Reptiles showed up about 300 million years ago, just east of Columbus, Ohio. The age of the dinosaur ranged from about 220 to 65 million years ago, or from about Pittsburg, Pennsylvania, 289 miles west of NYC to just 86 miles west of the Big Apple. Along the way, the first mammals showed up about 200 million years ago, early in the age of the dinosaurs, still near Pittsburg. Birds came on the scene about 150 million years ago, at about Harrisburg Pennsylvania on our toothpick path.

In tracing our toothpick path back from Spain so far,

an important point is that science is telling us that creaturely life has just been around for a small portion of time since the beginning of the cosmos, 600 million years out of 7.6 billion years. If we cut our 18,000 mile toothpick path into 12 equal parts, earthly life has existed only in the last one-twelfth.

It is important here to discuss the illusion of the status quo. Although as we look at the night sky, the stars and planets seem to behave predictably, it is important to know that stars are constantly dying and other stars are constantly being born. Gravitational forces cause collisions of asteroids, planets, stars and galaxies. We are propelled at unimaginable speeds by the expansion of the universe and the spinning of our own galaxy. In space there is no status quo. The story is the same on earth. As described above, the tectonic plates are moving and the earth's climate changes in dramatic cycles. We mentioned earlier that dinosaurs have come and gone. Scientists now believe that a long term climate change associate first with a large volcano and second with an asteroid impact caused an ice age that the great reptiles could not survive. The most recent ice age ended just 18,000 years ago long before we humans had any impact on the environment whatsoever. On earth there is no status quo. As we can chronicle that forms of life have come and gone for hundreds of millions of years, biologists tell us that new forms of life are still emerging while other forms become extinct. With life on earth there is no status quo. The universe is dynamic, the earth is dynamic, and life is dynamic.

Reconciliation with Genesis

This view of billions of years of unfolding creation

history has proven troublesome to many pious Christians who understand the prevailing notions of science to be at odds with the biblical creation account. To help relieve any such tension, the project next at hand is to see if we can reconcile the voice of science with the voice of God in scripture. The Genesis Chapter 1 account of creation continues the progression started at the beginning of this chapter. It is both beautiful and familiar. You can read it yourself in its entirety in just a few minutes and I encourage you to do so, starting at Genesis 1:1 and ending at Genesis 2:4. If you keep track of God's creative activity day-by-day in this set of passages, you might summarize the process something like this:

   Day 1: Light (and its separation from darkness)
   Day 2: Sky (dome in the midst of the waters)
   Day 3: Dry Land, the Earth, Seas and Vegetation
   Day 4: Sun, Moon, Stars (lights in the heavens)
   Day 5: Birds and Fish (creatures in the waters and the skies)
   Day 6: Animals and humankind
   Day 7: God rested

Immediately following the Genesis 1 account is another account in Genesis 2. It starts, "In the day that the Lord God made the earth and the heavens… then the Lord God formed man from the dust of the ground." (Genesis 2:4b-6) Although this account of creation is not as systematically presented as Genesis 1, that is, no day-by-day list of creative acts are presented, a close reading of Genesis 2 does yield an order in creative acts of God. Following is a simplified summary of this Genesis 2 order of creation:

First: Earth, heavens, male human
Next: Trees and vegetation for food
Next: Animals and birds
Next: Female human

Do you see any logical problems here? In the first account, just looking at it by itself, does it make sense that the earth came before the sun, moon and stars? How could plants come before the sun, especially since plants survive by photosynthesis? In the second account, does it make sense that male humans came before animals but females came after? Then between the two accounts there are some logical inconsistencies as well: did people come about before or after animals? Did earth come after the heavens or at the same time? These and a number of other issues, some of which will be described in the next section, must pique our interest in understanding in how these scriptures were written and what messages they were trying to convey.

Bible scholars don't just study the Bible; they also study a wide range of social, cultural and literary issues so that they can better understand what was going on in Bible times. They try to answer questions like, what were the social structures at the time? How did people live? What did they eat? Who were the foreigners among them and how did they live? Who were their enemies? What were the people's biggest concerns? Other than biblical scripture, what other written works existed at the time? As in science, the last century was a heyday for biblically-related archeological discoveries and insights. This wealth of new information is still growing and has not been totally digested.

In studying what other written works might have been known to the Hebrew people when Genesis was written, a Mesopotamian work was discovered that has both important similarities and differences with the Genesis creation accounts. Chiseled on seven tablets, the mythic legend of Enuma Elish[7] is believed to predate the writing of Genesis by 500 to 1000 years. In this strange mythic epic, creation resulted from a cosmic-scale bloody conflict between two gods, Timat and Marduk. The differences between Enuma Elish and Genesis are striking: a multiplicity of gods versus one God, creation cast as violent conflict versus an orderly process of creative love, a violent creation born from violence versus a creation that was seen by its creator as "good." Despite these differences, there are also striking similarities: A preexisting state of chaos, creation of divine light first, a day-to-day sequencing ending with a day of rest. Is this coincidence, or did a spiritually conscientious Hebrew sage attempt to articulate important truths about God using the outline of an already known story? In such a view, the Genesis writer introduces the important concepts of one God, creation from peace rather than violence, and a creation that is at once evaluated to be "good" rather than containing the stain of a bloody, violent beginning.

The process of thinking about what ideas a writer of scripture was really trying to convey might raise a broader question for you, how did Bible writers write? If we agree they wrote, somehow, from influence of the Holy Spirit which is the spirit of truth, the spirit of goodness, the spirit of integrity, we must also agree that they were, by all accounts, quite human themselves. This line of reasoning might lead to another question, is every Bible narrative intended to be factual? Certainly not. Sometimes the Bible writers made it easy to figure this out. For example, the Gospel writers were generally clear in telling us that the instructional stories that Jesus used to convey important messages are parables. In a parable, it can safely be assumed that the characters and plot elements are fictitious. But what about narratives in which the authors do not tell us? We must try to discover the author's intent and this is not always easy.

The Pontifical Biblical Commission's 1993 "Document on the Interpretation of the Bible in the Church"[8] highlights the importance of discerning the intent of the author in Bible scholarship. It encourages those who read the Bible to understand the literal sense of the writing as intended by the author, but expands understanding of the term *literal*. Under the umbrella of literal interpretation, the document carefully distinguishes among historical, metaphorical and allegorical intentions of the author.

> When it is a question of story, the literal sense does not necessarily imply belief that the facts recounted actually took place, for a story need not belong to the genre of history but be instead a work of imaginative fiction. The literal sense of scripture is that which has been expressed directly by the inspired human authors. Since it is the fruit of inspiration, it is also the sense intended by God, as principal author. One arrives at this sense by means of a careful analysis of the text, within its literary and historical context.[9]

This document goes on to underline the importance of historical research and the study of contemporary literary genres to fully understand the author's intent. This certainly opens the door for conscientious faithful to understand parts of the Bible as fictional when legitimate Bible scholarship so directs. It is reasonable to assume that such fictional stories still have a reason to be in the Bible and convey an important point. Such stories are allegories: they contain spiritual or moral truths conveyed in imagined situations or events. In allegories, the author generally intends to convey a message that is not specifically articulated in the narrative. In fact, the importance of an allegory can only be discovered behind the words.

Now let's return to creation. Let's say we accept the current scientific thinking about the process of the development of the universe. Let's also say we accept that the Book of Genesis was intended by its author to be allegorical, possibly even written as an adaptation of a pagan legend, but with important changes which conveyed important messages. Is there really any conflict? No, no conflict for Bible scholars, no conflict for the Church and no conflict for the laity.

Cardinal Joseph Ratzinger, now Pope Benedict XVI addressed this issue head-on in a series of four Lenten homilies delivered in the cathedral in Munich in 1981.[10] In these homilies, he broadly accepted the findings of science in his understanding of the timing and mechanism of the origins of the cosmos as possible. In so doing, he gives up no part of his faith and no part of the truth of scripture. Benedict emphasizes that the essential message from Genesis is that "God created", and that the exact mechanism by which this occurred is more a matter of curiosity than theological concern.

An important first task of this book is to reconcile God and science as we develop a view of creation history. In this first chapter we have outlined how faith and science can be rationally reconciled in talking about beginning of the cosmos and the development of life on earth. The important point here is to be open to the idea that God created; the startling and fascinating mechanics of how this came about are gradually unfolding.

Next we will take a step forward in creation and in time and think about the development of human life and our rise to humanity.

Chapter 3

The Path of Humanity

Then God said, "Let us make humankind in our own image, according to our likeness; and let them have dominion over the fish of the sea, and over the birds of the air, and over the cattle, and over all the wild animals of the earth, and over every creeping thing that creeps upon the earth. So God created humankind in his own image, in the image of God he created them; male and female he created them. God blessed them and God said to them, "Be fruitful and multiply, and fill the earth and subdue it; and have dominion over the fish of the sea and the birds of the air and over every living thing that moves upon the earth." God said, "See, I have given you every plant that is yielding seed that is upon the face of all the earth, and every tree with seed in its fruit; you shall have them for food. And to every beast of the earth and to every bird of the air, and to everything that creeps on the earth, everything that has the breath of life, I have given every green plant for food." And it was so. God saw everything that he had made, and indeed, it was very good. And there was evening and there was morning, the sixth day. Genesis 1:26-31

In the first chapter we explored the scientific view of the story the genesis of the cosmos, the earth and life. This set the stage for the beginning of the story of humanity. In this chapter we will explore the contemporary scientific ideas about the appearance of the human on the stage of the cosmos. Although the early part of this story is still somewhat sketchy, you will probably be surprised by the significant amount of new information and the implied complexity of the early part of the human story.

The Rise of the Human

Two branches of science help to sharpen the focus on how humanity developed. The first is anthropological archeology and the second is genome biology. Although at the time Darwin advanced his theory of evolution there was scant evidence of the presence of pre-humans, we are now rich with evidence of a wide array of our ancestors and their relatives. More and more physical examples of our ancient ancestors have been and are being found. Prehistoric bone fragments, bones, and even skeletons have been found which are definitely not from apes or monkeys and definitely not from humans like we are today. Our own specie, *homo sapiens*, seems to be a newcomer on the stage of the cosmos, around for just a couple hundred thousand years. Prior to the appearance of *homo sapiens* on the planet, there appear to have been a number of kinds of human-like species that came first. DNA mapping has opened the door to understanding how all of these DNA traces in these bits of bone fragments can be properly fit together into a mosaic of the story of the human family. Those who followed in Darwin's footsteps hoped to find a "missing link," a single pre-human example which bridged the evolutionary gap between ape and human. What has been found is much more complex and much more intriguing.

At this point in scientific study it appears that the first pre-humans showed up on the earthly stage about 6 to 7 million years ago. Like us, they walked on two feet, had some similarities in dental patterns and faced forward. Unlike us, their brain size and skeletal proportions were not quite the same as us today. Neither were they exactly ape or chimpanzee-like. They shared features that could be considered both human-like and features that could be considered ape-like. Of the dozens of different anthropological samples, literally hundreds of bones and bone fragments, DNA studies have determined that there were at least about 20 types, or families of human precursors. These were all probably unique species, that is, unable to have offspring with each other.[11]

Although it is strange to think of different species of humans, it is certainly not uncommon in the family of life on earth. It is not dissimilar to thinking of twenty different kinds of reptiles or mammals or birds. Some of the differences among the different kinds of human-like species were in size, shape and bone structure. These families, though they had discernable differences, can be rightly thought of as different branches of a family tree. They were all related but not all the same. Only one such family, one branch, from which all living humans appear to have come, remains. It has been suggested that our oldest common ancestors can be traced back at least 1.8 million years, to a specimen named *homo ergaster*. It may soon be proven that a specimen 3.2 million years old, *A. afarensis*, is a common human ancestor. This is the family of the widely toured "Lucy" skeleton found in Ethiopia in the 1970's.[12] The hundreds of bones found to date are like shapshots in time, like a variety of photos on the family bulletin board. Scientists are still developing the relationships among the family members.

Moving on to focus on our immediate family, *homo sapiens*, DNA studies are casting a new bright light on the story of the human past. Such studies now allow scientists to determine biological family links. Just as a DNA study can link you to your mother and grandmother and so on, genetic signatures can determine if a person has any kind of relation with another. This process allows scientists to trace backwards common gene signatures from generation to generation. Using this process, scientists believe they have found a common denominator of the existing *homo sapiens* family, possibly a female lineage with whom we all seem to share a genetic marker. This lineage can be traced back to about 150,000 years ago. Not only can we trace the time, but not unlike tracing back the movement of the stars to the big bang, scientists can trace back our ancestor's migration patterns with studies of mitochondrial DNA to a physical place. The place seems to be eastern Africa, probably in the broad neighborhood of today's Ethiopia.[13] Work in these fields continues at a robust pace and more fascinating surprises seem likely.

Having listened to the voice of science to trace the story back to its beginning, at least as best we understand the human beginning today, we can also trace it forward to get an idea of when human migration occurred. It appears that this east Africa cradle of humanity was a good home for our ancestors except during periods of climate change. Scientists speculate that people followed wandering herds within the general area except when climate changes demanded that food be sought elsewhere. In related studies, scientists who study the history of the earth's climate patterns confirm this anthropological view. It appears that there were a number of such migrations, including migrations by human-like species which predate our own. Within our own specie, *homo sapiens*, there apparently was a first small migration about 60,000 years ago, and then a much larger migration that continued from 40,000 to 20,000 years ago. It is this migration period which the first of our ancestors moved to Europe, Asia and across the Alaskan land bridge into the Americas.[14]

Although we have sometimes thought about people of these times as having lived what we might consider savage lives, archeological finds indicate that our ancestors of this migrating age were more socially advanced and interconnected than we might have previously guessed. Although they generally seemed to set up settlements where food was plentiful, set up settlements they did. There is evidence that trade routes existed as far back as 20,000 years ago and perhaps even farther. Agriculture seems to have been pioneered out of necessity in the Middle East during a dry climatic period about 8,000 years ago. Farming practices were then disseminated, along with a variety of hand-produced goods, among widely geographically diverse populations, ostensibly by both land and water routes.[15]

Now let's return to our toothpick path to put these time frames into perspective. Based on the anthropological finds in Africa, our earliest know pre-human ancestors are believed to have preceded us by about 6 million years. Six million years in toothpicks? Only about 8 miles. Manhattan Island is about 13 ½ miles long and about 2 ½ miles wide, so the pre-human/human toothpick path would extend a little more than half the length of Manhattan. It should be pointed out that that this eight-mile path begins with human-*like* ancestors, but anthropologists tell us that humans with social tendencies and cognitive abilities that we would identify with being human didn't come on the scene until about 100,000 to 200,000 years ago. Considering humans with social and thinking skill similar to our own, our path is shortened to less than one-quarter of a mile in toothpicks, only about one and one-half city blocks long in New York!

If your imagination has been engaged in thinking about all of these silly toothpick paths, you have already been struck by an important insight. Compared with the 18,000 mile path of the cosmos or even the 6,000 mile path of the earth, the human path of less than a quarter of a mile, less than two blocks long, is insignificantly small. Recall that creaturely life has existed on earth for only about one-twelfth the time that the cosmos has existed. Human life has existed for about one-three thousandth of this span of creaturely life. We humans have been here for only a moment in the vast eons of time. In the life of the cosmos, we have been here for only a few breaths, a few heartbeats, a few blinks of an eye.

Although anthropologists may consider our modern-like human ancestors to have been around for as much as a couple hundred thousand years, are there any other important distinctions in stages of human development? There are a number of such distinctions, but the distinction between prehistory and history is most notable, reflecting a society's development of language and writing skills along with a degree of grasp of self-knowledge. One of the earliest hinge points of pre-history to history was in Egypt, with this hinge-point generally accepted to be about 3200 B.C.E. At this point, language, agriculture and architecture had been developed well beyond independent family-tribal patterns.[16]

In toothpicks? We must now shift our scale from miles to feet: just 22 feet. This portion of the toothpick path which starts at the center of the alter and extends down the center aisle ends inside St. Patrick's Cathedral, not just inside, but barely to the first pew. Now our contrast in cosmos-time and human-time is even more startling. The comparison of our toothpick path analogy is 18,000 miles versus 22 feet of recorded history. We civilized humans have barely taken a few steps on the stage of the cosmos!

The description of humanity to this point focuses on the external, what a trans-historic observer might have seen through long range binoculars. But is there an internal element to consider as well? Do the biological roots that we came from influence us today? Do they influence us in our biological processes only or in our behaviors as well? These questions open our reconciliation to the fields of medicine and psychology.

Scientists who map DNA have found that human DNA is not as different from other forms of life as we might have suspected. For example, we share about 98 percent of the same DNA coding with our closest relatives in the animal kingdom, the chimpanzee. Accepting that there is an undeniable resemblance between human DNA and the DNA of some other mammals, it is not difficult to believe that we have some of the same instincts as early humans or even animals. The concepts of survival of the fittest, drive to reproduce, of storing up food and resources for hard times, and of protecting the family unit at all costs may be easily seen as concepts which may have been passed down to us through our DNA. In fact, these instincts may have been the controlling factors of our pre-human ancestors for millions of years as they slowly increased in skills and knowledge and learned to take control of their surroundings rather than allowing their surroundings to control them. In this long initial period, our ancestors probably stuck close to their families and tribes and defended their families and territories against the threats of animals and other humans, and learned to increasingly protect themselves from natural disasters.

The view that we are burdened by some leftover biological equipment and instinct from the past is becoming prevalent in public consciousness. A You Tube video can explain that the seemingly useless human appendix was once critical in digesting a more cellulose-based diet. In the bestselling series of self-help book dealing with health improvement, Drs. Michael Roizen and Mehmet Oz incorporate the theme of leftover traits into their thinking about some of our contemporary problems.[17] For example, on the topic of increasing trends of obesity in the U.S., they point out that part of the problem is that we generally have an overabundant supply of food and following our instincts to hoard in times of plenty, we tend to heed a biological call to hoard food in our bodies despite the call of our doctors and our egos to eat more responsibly. Drs. Roizen and Oz also point out that foods new to the human diet including super-calorie-packed sweets and prepared foods tend to trick our minds in judging how much food is enough.

Similarly, our society's focus on physical beauty and increasingly graphic depictions of sex in the media could be an overreaction to a quite natural and quite strong drive to procreate. Such a strong drive might have been totally appropriate in a world with only a smattering of people in which huge challenges had to be met to successfully rear a family. However, with the same strong drive in a much different society, biological drives could be seen as contributors to societal problems: unwanted pregnancies, the spreading of sexually transmitted diseases and increasing levels of associated crime. Drs. Roizen and Oz refer to some of these human internal drives to be *hard-wired*, that is, at work on an interior level of our mind that we do not access by our conscious mind thought patterns. They consider the drives for sufficient food, drink, sleep and love/sex to be in this category.[18] They note that our responses to insufficiencies in these areas sometimes become extreme, far beyond the range of normal rational behavior.

Likewise, a contributing factor to violence in contemporary society could stem from the natural instinct to protect one's family and tribe from the danger of death from other violent animals and tribes. In this view, violent instincts once perfectly natural and perfectly necessary for human survival can be misapplied in non life threatening situations. If not adapted properly to the current human condition, unchecked instincts can prove deadly.

To this point in our discussion of humanity we have not discussed religion or notions about God. The story of the human-God relationship surely begins in prehistory, but we can only speculate about what our early ancestors experienced and believed. We can speculate that early humanity felt relatively powerless when the forces of nature turned against them in storms, floods, earthquakes, volcanoes, tornadoes and the like. We can imagine early people peering out of their caves or huts during such natural phenomenon and wondering whether their family could survive such a force. At some time in this process of considering their helplessness, an intuition might have developed that some power existed beyond them. Somehow in a number of cultures in a number of ways over just the last tens of thousands of years, the idea of some transcendent beings in charge of these natural phenomena developed. These are the mythologies of various cultures. These mythologies reflect numerous gods, often depicted as more arbitrary and capricious than the humans who imagined them. Such imaginative mythologies developed in essentially every human culture.[19]

Genesis and the Human

A literal regression of genealogies in the Bible would date Adam and Eve's emergence onto the earthly scene in about 6000 B.C.E. The voice of science tells us that humans much like us have been around 20 times that long and that pre-humans were on the scene millions of years before. There are two very simple options in resolving the difference which make further analysis unnecessary. First, one could believe that the Bible is literally true and that science is sadly mistaken, hoodwinked by a lack of understanding of the mysterious creative mechanisms of God. Second, one could take the opposite view that science has made its case and the biblical accounts are without merit. Although either of these options presents an easy way out of the dilemma, scientifically-oriented theologians seek a more nuanced way of understanding how both science and Bible can be true. The key to unlocking the seemingly irreconcilable views of science and Bible, in particular the Old Testament, is once again to accept that the Bible is not intended to be a primer on physics or anthropology.

Although the Bible contains some historically correct information, it also contains some stories which were intended to convey theological truths.

Returning to the two accounts of the creation of humanity in Genesis 1 and 2, how can we reconcile them with each other and how can we reconcile them with the scientific view?

In trying to reconcile the two creation accounts with each other, it is helpful to think more about what the sacred scriptures are and how they were put together. Bible scholars think that many of the narratives, particularly in the Old Testament scriptures, existed in oral tradition before they were committed to writing. This is especially true for the historically earliest events depicted in the Bible, many of which ostensibly happened before writing existed. We might imagine early generations of the Hebrew people sitting around a fire and seeking answers to the same questions which engage us today. How did the earth and stars come to be? Perhaps the Hebrew sages passed down the Genesis 1 narrative from generation to generation to explain creation by a single God, a loving God, in language their families could understand. Around other fires, perhaps there was another question: why is there evil? Perhaps those Hebrew sages passed down the Genesis 2 narrative to explain that ignoring God leads to separation from God and that personal selfishness results in conflict and unhappiness. Perhaps, as was suggested in the first chapter, these narratives were sometimes adaptations of other well-know narratives from other cultures, but changed in important ways to reflect precepts of the Hebrew faith.

Then there is the question of the actual writing. Bible

scholars have meticulously studied the Old Testament manuscripts word by word comparing the varying use of words and phrases. In the case of Genesis, scholars conclude that there were probably several contributing authors who brought with them different oral traditions. The works of these authors were later assembled by another writer, essentially an editor, and additional editorial smoothing may have occurred at other points of scriptural development. These ideas are neither new nor revolutionary. The introduction to Genesis in the 1970 St. Joseph's Version of the New American Bible summarizes this idea of multiple authors well: "Despite its unity of plan and purpose, the book is a complex work, not to be attributed to a single original author. Several sources, or literary traditions, that the final redactor used in his composition are discernable. These are the Yahwist (J), Elohist (E) and Priestly (P) sources, which in turn reflect older oral traditions."[20]

Once again, we are not forced to choose between science and the Bible. Rather, we are challenged to search out the root messages that the inspired authors hoped to pass down from generation to generation. For example, this chapter begins with a passage from the first biblical creation account, Genesis 1: 26-31. In the first sentence the reader is told that humankind is created in God's image. This theme is repeated twice in succeeding sentences so it must have had importance in the mind of the author. Although we can't be sure what was in the sacred writer's mind's eye, we do know that in the writer's time the same word for *image* was used in another sense. In the empires of antiquity, in the corners of the empire far beyond the personal presence of the emperor himself, a statue of the emperor was erected as a symbol of his power and presence. Is this what the sacred writer had in mind?[21] Are we humans God's representatives on earth, intended to be symbols of his presence and power? Are we intended to be God's representatives in creation? This idea resonates with me and with countless others.

Also in this first creation account in Genesis, both male and female are created together. And, more importantly, humankind is presented as the capstone of God's creative process. The verses of Genesis preceding this verse describe God's other acts of creation day-by-day. In every other case, the heavens, the earth, the sky, the stars, the birds and fish, God deemed them as "good." But after creating humankind and assessing his work, God deemed his creation of humankind "very good." In this first account in Genesis, God seemingly saves the best for last, creates humankind in his own image, and is particularly pleased with his work in creating humanity. "God saw everything he had made and indeed, it was very good." This can be understood as a historical human foundation for the dignity of human life.

In the story of humanity presented in this chapter, we have seen that the emergence of the human from the animal family was likely a slow and complex process with many dead-ends on the branches of the family tree. Our branch alone has survived, and we have populated the earth. In this process we have developed tools and languages and have learned to exert some control over the threats of natural forces that once determined our fate. In this growth we sensed a transcendent force and sometimes attributed to this force human tendency for lust and violence. Then, starting about 800 B.C.E., a funny thing happened. Against the backdrop of multiple all-too-human gods and human individuals and societies operating on the animal instincts of self preservation and hoarding, something different happened. This change will be the topic of our next chapter.

Chapter 4

The Axial Age

Happy are those who do not follow the counsel of the wicked,
Nor go the ways of sinners, nor sit in the company of scoffers.
Rather, the law of the Lord is their joy; God's law they study day and night.
They are like a tree planted near streams of water, that yield its fruit in season;
Its leaves never wither; whatever they do prospers.

But not the wicked! They are like chaff driven by the wind.
Therefore the wicked will not survive judgment,
nor will sinners in the assembly of the just.
The Lord watches over the way of the just,
But the way of the wicked leads to ruin.
Psalm 1

With a view of humanity developing over the long haul of time there were progressive agricultural, architectural, social and language achievements. These were all important aspects of the blossoming of humanity. Volumes have been written about these aspects of many different civilizations around the world including the distinction between history and pre-history. It is fair to say that the timing of the pre-history to history hinge has been found to vary widely among civilizations. Most of the criteria for evaluating civilizations relate to objective achievements, but anthropologists recognize other important aspects of evolving human behavior: moral and spiritual elements.

In the analysis of the development of major world civilizations it has been noted that within the range of the same thousand years, just a single heartbeat on the cosmic scale of time, civilizations which were independent of each other developed systems of beliefs and behaviors which transformed their societies. These systems of thought became the philosophical and spiritual underpinnings of not only ethical human behavior, but also the beginnings of the major religions of the world today. Since the changes of that time are considered so profound, akin to a new moral and spiritual awakening for humankind, the period has been singled out by some scholars as the most important time of thought development in human history. It is called the *Axial Age*.

Both the idea of the Axial Age and the term itself are credited to twentieth century German philosopher Karl Jaspers who used it to describe the period from roughly 800 B.C.E to 200 B.C.E.[22] According to his view, similar revolutionary thinking appeared in China, India, the Middle East and Ancient Greece at about the same time. Jaspers observed that the spiritual foundations of humanity were laid simultaneously and independently, and that these foundations still undergird our religions and highest ethical beliefs today.

In China, both Confucianism and Taoism had their roots in this period. In India, Hinduism was established and Siddhartha Gautama, or the Buddha, also lived during this period. In Persia, Zarathustra's life gave rise to

Zoroastrianism. In Greece, Plato, Socrates and Aristotle, among others, laid philosophical foundations which were incorporated in the movement to monotheism in general and later into Christian theology in particular. In Israel, the rise of the prophets in this period promoted the radical notions of both monotheism and helping the underprivileged in society.

In short, this was a time in which humanity in its civilized centers independently seemed to have received the same wake up call, ostensibly from the same source. Although manifested in much different forms in different places, the underlying themes of transcendence and ethical behavior toward fellow humans had remarkable similarities. Following are several very brief overviews of some of these movements. Although these sketches are individually incomplete in describing the main figures and belief systems, the purpose here is draw out some common elements in the fruit of the movements.

Judaism

What was going on in Israel in and around the period Jaspers refers to as the Axial Age? Plenty. The Axial Age opens at the end of the glory days of the Kingdom of Israel. David was on the throne from about 1000 B.C.E until 960. After his son Solomon's reign the kingdom was split and both kingdoms were subsequently conquered. The Northern Kingdom was conquered by Assyria in 722 B.C.E. The southern kingdom of Judah fell to Babylon with the destruction of Jerusalem in 586 B.C.E. and the Babylonian exile spanned the years 587 to 538 B.C.E. Although the nation was split, vanquished, and significant portions of the population were taken as slaves to Babylon, the Axial Age was still extraordinarily fruitful for the people of Israel from a theological standpoint. Although the history of the Hebrew people is traced back in the Bible as much as 3000 years before the Axial Age, it is during the Axial Age when many of the Hebrew scriptures were given their form. It is not surprising that the written works of prophets who lived in the axial timeframe became part of the Jewish scriptural tradition, but some works once thought to be written much earlier are also now seen as axial products. It is possible that these accounts of events which occurred much earlier had been passed down by oral tradition and first committed to writing during this period. It is possible that they preexisted but were edited in this period. It is possible that some were editorial collections of various stories from various sources. Although Bible scholars differ on exactly what material was new and what was simply adapted in this period, there is a growing body of scholarship that indicates that a large volume of Old Testament scripture reached its final form in the Axial Age.[23]

Numerous examples of Axial Age-theme lessons are present in the Old Testament. The prophetic books are particularly rich in this new definition of virtue, linking virtuous living with what it means to serve God. For example, this is one of the prominent themes of the prophet Amos. In Chapter 8, God tells Amos that he is at the end of his rope in tolerating injustice among the well to do of Israel. The following excerpts highlight this point, but the entire chapter is well worth reading:

> The time is ripe to have done with my people Israel; I will forgive them no lon-

> ger. Hear this, you who trample on the needy and destroy the poor of the land! I will turn your feasts into mourning and all your songs into lamentation. I will cover the loins of all with sackcloth and make every head bald. I will make them mourn as for an only son, and bring their day to a bitter end. (Amos 8: 2b, 4, 10)

Axial Age themes are also plentiful in the Book of Isaiah. The theme of social justice permeates the book. In Chapter 58 the author of the final portion of Isaiah emphasizes that God is much less interested in ritual than "sharing your bread with the hungry, sheltering the oppressed and the homeless; clothing the naked when you see them, and not turning your back on your own." (Isaiah 58: 7) Although much of the Old Testament tells the story of the relationship between Yahweh and the Hebrew people, the author of the last portion of Isaiah expands God's concern to the just foreigner as well:

> Thus says the Lord: Observe what is right, do what is just; for my salvation is about to come, my justice, about to be revealed. Let not the foreigner say, when he would join himself to the Lord, 'The Lord will surely exclude me from his people;' All who keep the Sabbath free from profanation and hold to my covenant, them I will bring to my holy mountain and make joyful in my house of prayer; for my house shall be called a house of prayer for all people. (Isaiah 56: 1, 3a, 6b, 7a, 7c)

The Psalms, also generally Axial Age products, are rich in the theme of virtue. Psalm 1 is presented in its entirety as the introduction to this chapter. In this psalm, the ways of the just are contrasted with the ways of the wicked. This contrast is a major theme of the Psalms; admonitions against the "wicked" appear in 19 Psalms.

In the Old Testament wisdom literature, particularly the book of Proverbs, wisdom is broadly correlated with justice and righteousness and contrasted with wickedness. Although there was no specific word for "virtue" in the Hebrew language, the Axial Age teachings of the Old Testament are unmistakable in tone and intent.

Hinduism

Hinduism has roots that predate the Axial Age by 2500 years. Rather than a single movement that a single name implies, practice has varied significantly in different areas of India and in different times. During the Axial Age, however, some important teachings were assembled which still are seen as foundational in Hinduism today. Consistent with the widespread moral awakening of humanity in the Axial Age, the Upanishads teach lessons in compassion and virtue. Most forms of Hinduism embrace a transcendent Spirit and an internal path present in each person to access the transcendent. Coupled with a life of virtue, this connection with the transcendent, what we would call God, is a chief aim: "He who sees the meaning of this Upanishad and has cast off all wickedness stands in the Infinite, Supreme plane of heaven."[24]

The Hindu focus on the internal life, seeking God within, gained importance in the Axial Age. Although the impulse is expressed much differently in the Judeo-

Christian tradition and Greek thought, the aspects of introspection and connection to God through prayer seem to be legitimate common elements.

Buddhism

Siddhattha Gotama, latter known as the Buddha, the awakened one, was probably born about 490 B.C.E. Although he was born into a royal family, at the age of 29 he left behind his life of power and comfort to seek wisdom. In particular he sought out to understand how humans could be happy with suffering ever-present in human life; he sought freedom from suffering. He followed various practices of the sages of his time including yogic meditation and self-mortification which he ultimately found unsatisfactory. He achieved a life changing transformation after devising a *middle way* in which he embraced both physical well-being and meditation. His transformation, self-described as *awakening*, resulted in a freedom from suffering and mental attachments.

Some aspects of Buddhism seem so foreign to the Christian mind that one might question the value of including this belief system in dialogue about progressive human development. In particular, the commonly discussed tenets of no God, no soul and rebirth might make Buddhism seem an odd source from which to mine developing human wisdom. A deeper understanding, though, underscores the significance of practice over dogma; prayer over human intellect. Further, a more nuanced understanding of the Buddhist view toward the transcendent is neither acceptance nor rejection of God or soul. The awakening experience is not easily summarized, but the idea of a transcendent connection with timeless creation might be considered an over-arching aspect.

Although this brief sketch of Buddhist spirituality and beliefs is far from complete, the more germane aspect of Buddhism for this study is its fruit. These fruits are compassion for all humans and all forms of life, non-violence, and life lived to alleviate the suffering of others.

Like other Axial Age wisdom figures, Buddha emphasized virtue, sometimes expressed as goodness: "A man should hasten towards the good, and should keep his thoughts away from evil; if a man does what is good slothfully, his mind delights in evil."[25] "Follow the law of virtue; do not follow that of sin. The virtuous rests in bliss in this world and the next."[26]

Buddhism was from the beginning against the wielding of power by violence. These words against violence are attributed to Buddha: "A man is not just if he carries a matter by violence; no, he who distinguishes both right and wrong, who is learned and guides others, not by violence, but by the same law, being a guardian of the law and intelligent, he is called just."[27]

These are Axial Age values and they remain cornerstones of Buddhist belief and practice today. This may have been brought to mind in 2008 when the international press documented the peaceful demonstrations of Buddhist monks against a corrupt regime in Myanmar.

Confucianism and Daoism

Prior to the Axial Age, China shared many attributes with the more familiar Mediterranean world. Family tribes and states warred for control of people and natural resources. Leaders sought the help of various gods to support them in their conquests. As the Axial Age dawned, a new group, the Zhou dynasty came to power in about 1045 B.C.E. It was in this dynasty that virtue among leaders and the people alike became a discernable theme. Although this dynasty did not survive the challenges of war throughout the axial period, the idea of the importance of moral living did persevere and provide the backdrop for the rise of two important Axial Age figures.

Confucius was born about 551 B.C.E. Interestingly, this name was later coined by Jesuit priests in the 17th century. He is known in classical Chinese literature as Kongzi, or Master Kong. Born into lower ranks of nobility, Confucius was conscious of the role of leadership in setting appropriate examples to achieve social justice and harmony. He further developed the theme that a leader's legitimacy is contingent on virtue, and that a just society can only be realized under the guidance of just rulers. He was passionate about the arts, particularly music, and felt that arts added to virtue and harmony in human life. His version of the golden rule was based on trying to ascertain the likely desires of other people and acting in accord with their wishes. His teachings embrace a profound empathy towards others that Christians could easily compare with Christ's command to love ones' neighbor. Although there is little religious ritual in Confucianism, he did encourage a prayer-like practice of silent sitting. Confucius had little to say about gods or souls, but his teachings of moral and ethical precepts gained so broad a following that the movement transcended his own time and government.[28]

A second important Chinese Axial Age figure is Laozi, founder of Daoism. Laozi was a contemporary of Confucius. Dao, the way, refers to the way of nature, the way of people and the way of the ineffable. In contrast with Confucius, who was most interested in aspects of culture, Laozi was more interested in harmony with nature. Through meditative prayer, the Dao way taught that human harmony could be achieved not through a system of moral rules, but by an inner conversion which resulted in profound care for others. "The sage is constantly good at assisting the people, he does not abandon them."[29]

Daoism also shares distaste for war and violence with other Axial Age movements. Regarding weapons, Laozi teaches, "Weapons are the tools of ill omens. Do not regard them with delight. To regard them with delight--- this is to enjoy killing people. When a war is won, the occasion is treated with funeral rites."[30]

On a deeper level, the Dao emptying of self is akin to the *kenosis* of Jesus referred to in Christian Gospels. Another important concept in Daoism is *wu wei*, sometimes described as actionless action or powerless wielding of power. Although difficult to describe, it suggests that the duality of acting and not acting is somehow transcended when in harmony with creation. Although there were further ritual and experiential developments of Daoism after the Axial Age, the lessons of humility, empathy and peace throughout social structure endure.

Philosophical Rationalism: the Greeks

Greek thought in the Axial Age was truly revolutionary. As in the other civilizations of the Axial Age, the new thinking encompassed a wide array of topics. These topics ranged from mathematics to forms of government. The concept of the democracy was born in this age. Not only were a wide variety of individual topics explored, but a focus on reason and inquiry gave humanity a new framework for developing knowledge.

There were a number of important Greek figures involved in the many different aspects of this movement. Probably the three most well known are Socrates, Plato, and Aristotle. Socrates (469-399 B.C.E.) emphasized the importance of virtue in the human life. His well know instruction to "know thyself" raised the level of human conscience and consciousness by highlighting the importance of constantly analyzing one's own motives and actions. His teaching of virtue was not simplistic, but nuanced in dealing with difficult moral situations of his time. For example, in analyzing the duty of a soldier in the Greek military, he introduced the idea that it is correct to disobey an immoral command. His focus on the importance of virtue is summed up in his statement, "it is not living that matters, but living rightly." Socrates examined the eye-for-an-eye and tooth-for-a-tooth mentality of his day and developed are more nuanced ethical stance: "One who is injured ought not return the injury, for on no account can it be right to do an injustice; and it is not right to return an injury, or to do evil to any man, how ever much we have suffered from him."

Socrates' student Plato (427-347 B.C.E.) carried on the work and expanded it considerably, most notably in the areas of acquiring knowledge, ethics and justice. His several dozen "Dialogues" which consist of long series of questions and answers, show how difficult issues can be systematically explored. In regard to society as a whole, he underscored the importance of education in producing a just society. Plato was remarkably ahead of his time in regard to human rights: "All men are by nature equal, made all of the same earth by one Workman; and however we deceive ourselves, as dear unto God is the poor peasant as the mighty prince."

Plato's student Aristotle (384-322 B.C.E.) again took his predecessor's teaching to another level. One of his greatest contributions to humanity was his focus on empiricism. He taught that by trying out new ideas then carefully observing and analyzing results, new knowledge could be gained. This, of course, is the foundation of the scientific method. He followed in the footsteps of his predecessors by advancing the concept of human virtue. In summarizing his thoughts on virtue, he said, "all virtue is summed up in dealing justly." Recognizing the human potential for both good and evil acts he said, "At his best, man is noblest of all animals; separated from law and justice, he is the worst."

The influence the many Greek philosophers in the Axial Age cannot be overstated. Their ideas were built upon in Christian and Muslim theological and intellectual movements. In the Christian tradition, for example, St. Thomas Aquinas brilliantly used the methodology and logic of Aristotle in his *Summa Theologica* in which he developed a complete Christian theology including a series of logical proofs for the existence of God. In addition to developing new ideas about virtue, justice, and human rights, the Greek influence looms large for all of Western Civilization in the institution of democracy and the rise of science.

Axial Age Common Denominators

Each of the Axial Age religious movements had different religious traditions and cultures to assimilate and adapt, so it is not surprising that their dogma and ritual varied widely. In fact, these vastly different cultures from which the axial movements began were so different that it is remarkable that there were any common denominators at all. Within these new movements which were the seeds of our current major world religions, there appears to have been a common spirit. This spirit gave rise to the consideration of actions beyond self-interest. These movements all taught that a person should look beyond satisfying biological need and advancing the interest of tribe. These movements all differentiated between what a person could do and what a person should do. It was a beginning of systematic human moral thinking, in defining what it means to be human. Each tradition developed its own version of golden rule, but all had the same core message: not doing to others what you would not have done to you.

It was as if at this time in history humanity was given a wake-up call to look beyond personal and tribal interests. Humanity was taught that might doesn't always make right. People were taught that something exists beyond what we can see and that the human and the transcendent can commune.

Following is a list of some of the morally-inclined ideas of the Axial Age. Although these ideas were probably individually present earlier, in the Axial Age they were given important voice, systematized, and preserved in such a way that following generations would be influenced.

1. Compassion in the presence of suffering: Although suffering is part of human life, it should be alleviated as much as possible for all people.
2. Thoughtfulness is superior to brute force: Analy-

sis of one's situation, capabilities, intentions and possibilities should precede taking action.
3. Transcendence in the core of the human: There is some superior force/personality/intelligence that can be accessed by the human by habitual practice of contemplation or prayer.
4. Connectedness: Through contemplative practice, the connectedness of all of creation can be sensed.
5. Mindfulness of needs of others: Out of connectedness and compassion, one senses the needs of others and should respond appropriately.
6. Positive notions about structure of society: Just as a person's capabilities, actions and intentions can be analyzed; one's society should be analyzed and changed for the better.
7. Against hatred, greed, egoism: The opposite of positive moral values, those characteristics which inhibit application of compassion, should be avoided.
8. Against violence: Violence on large and small scales was seen to cause suffering. The idea was advanced that issues should be resolved without applying power by violence.

Although these eight factors have overlap, as a group they convey the idea that a new system of thought about virtue and human morality emerged during the Axial Age. This permanently raised the moral bar for humanity.

Although Jasper's ending of the period in 200 B.C.E. excludes further important developments in world religions, namely Christianity and Islam, the teachings of Jesus and Muhammad were in harmony with axial themes. How these teachings harmonized with both axial ideas and the issues of their times will be discussed in following chapters as we follow the toothpick path of time.

Jasper's 600 year duration of the Axial Age represents about four feet of toothpicks in total. Those four feet occurred from 2200-2800 years ago. Along our toothpick path, that's 15 to 19 feet of toothpicks ago. In St. Patrick's Cathedral, the four feet of the Axial Age toothpick path which began at the center of the altar would lie near the first pew.

An important concluding point to make about the Axial Age is that the roots of all of the major world religions, and many of the minor religions as well, are accounted for in the Axial Age. Including Christianity and Islam, both of which sprang from Hebrew Axial Age roots, about 85 percent of humans today belong to a religious tradition with Axial Age beginnings.

Chapter 5

Jesus

In the beginning was the Word,
  And the Word was with God,
  And the Word was God.
He was in the beginning with God.
All things came to be through him,
  And without him nothing came to be.
What came to be through him was life,
  And this life was the light of the human race;
The light that shines in the darkness,
  And the darkness has not overcome it. John 1: 1-5

Jesus was born just past the end of what has come to be known as the Axial Age. Positioning him outside of the group of axial wisdom figures might seem to the Christian as though he is being denied a rightful place among great teachers. In another way, the timing might be seen as Providential. Jesus came at a time when the then civilized world had already started coming to grips with Axial Age teachings. Among the Hebrews, the moral lessons of the Torah were widely known and practiced, although practiced in different ways by different groups of Jews. Perhaps in Jesus' teaching we can see that the first set of lessons in humanity had been learned, at least in a superficial way, but that there was a deeper lesson to be taught and the time was right. Jesus' moral teachings took

Axial Age teachings to a new level showing a way forward for humanity.[31]

In thinking about the lessons of Jesus it is helpful for purposes of this study to distinguish two aspects of Christ. The first aspect is that of the Incarnation, Jesus fully God and fully human, fully destined for his mission of salvation. This aspect is accepted as foundational in Christian faith. Second is the aspect of Jesus as extraordinary moral teacher, an aspect with which most of humanity today agrees. Muslims, Buddhists and Hindus all agree that Jesus was an extraordinary wisdom figure and that his moral teachings were revolutionary and are profoundly important. Since this book deals primarily with humanity as a whole and explores the possibility of humanity on a path of moral progress, Jesus' introductions of new standards in moral teaching are of special interest.

The task at hand, then, is to identify Jesus' moral lessons that challenged human thought and behavior in a new way, beyond axial ideas. The Gospel of Matthew makes this an easy endeavor. The Sermon on the Mount, in particular, is ground-breaking in two respects. First is that the prevailing notion of who is blessed by God is shattered by Jesus in the Beatitudes. Second, Jesus teaches that the commandments of the Torah are just a primer and that there is a new, more demanding standard.[32]

In the Jewish view of Jesus' time, living in accord with the laws of Moses was expected to be rewarded in the present by God. Generally, those who had comfortable, reasonably healthy lives were deemed to have been living within the law and blessed accordingly. Those afflicted with poverty or chronic illness were believed to have sinned, or even to have had a near ancestor who had sinned. This general notion is evidenced as the over-arching theme in the book of Job: Why did God subject Job, an upright person, to loss and humiliation? In the Beatitudes, Jesus explained that the old notion of who is blessed and who isn't is far from correct.

The gospel writers have passed down two versions to us, one in the Gospel of Luke referred to as the Sermon on the Plain and one in the Gospel of Matthew referred to as the Sermon on the Mount. Although Matthew's version is more familiar, we will look first at Luke's version, Luke 6: 20-22:

> Blessed are you poor, for the kingdom of God is yours.
> Blessed are you who are now hungry, for you will be satisfied.
> Blessed are you who are now weeping, for you will laugh.
> Blessed are you when people hate you, and when they exclude and insult you,
> > And denounce your name as evil on account of the Son of Man.

Luke follows the blessings with corresponding woes, Luke 6: 24-26:

> But woe to you who are rich, for you have received your consolation.
> But woe to you who are filled now, for you will be hungry.
> Woe to you who laugh now, for you will grieve and weep.

> Woe to you when all speak well of you, for
> their ancestors treated the false
> prophets in this way.

In this address, Jesus seeks to turn the value system of his people, the Jews, upside-down. Wealth and success no longer reflect God's favor, rather God's warning of an impending reversal of fortune. For the socially disadvantaged, Jesus predicts a reversal also, from homelessness, hunger and sadness to a life of fullness and joy.

The more familiar version of the Beatitudes from Matthew's Gospel are less anchored to physical status of hunger and poverty, but rather deal with more interior aspects of humility, integrity and compassion:

> Blessed are the poor in spirit, for theirs is
> the kingdom of heaven.
> Blessed are they who mourn, for they will
> be comforted.
> Blessed are the meek, for they will inherit
> the land.
> Blessed are they who hunger and thirst
> for righteousness,
> For they will be satisfied.
> Blessed are the merciful, for they will be
> shown mercy.
> Blessed are the clean of heart, for they will
> see God.
> Blessed are the peacemakers, for they will
> be called children of God.
> Blessed are those who are persecuted for
> the sake of righteousness,
> For theirs is the kingdom of heaven.

> Blessed are you when they insult you and persecute you and utter every kind of evil against you falsely because of me. Rejoice and be glad, for your reward will be great in heaven. Thus they persecuted the prophets who were before you.
> Matthew 5: 3-12

It is clear that Matthew's version is even more profound in teaching the necessity of interior transformation. Reading the lines together seems to imply that attitudes of humility coupled with sincere desire for holiness result in a life of mercy and peacemaking, but not without personal cost. In the field of moral theology, the Beatitudes are often considered the benchmark of human intentions and endeavors.

Also in the Sermon on the Mount, Jesus takes the Ten Commandments to a higher level. He introduces several of the commandments and other aspects of Mosaic Law with the phrase "You have heard that it was said..." He then takes the meaning to a deeper, more internal, more challenging level beginning with the words, "but I say to you..." Following is an example dealing with enmity:

> You have heard that it was said, 'You shall love your neighbor and hate your enemy.' But I say to you, Love your enemies and pray for those who persecute you, so that you may be children of your Father in heaven; for he makes the sun rise on the evil and on the good, and sends rain on the righteous and the unrighteous. For if you love those who love you, what reward

> do you have? Do not even the tax collectors do the same? And if you greet only your brothers and sisters, what more are you doing than others? Do not even the Gentiles do the same? Be perfect, therefore, even as your heavenly Father is perfect. Matthew 5:43-48

The command to love one's enemies is, without doubt, a profound step beyond the Mosaic Law to love one's neighbor. In this same set of passages, Matthew 5: 21-48, Jesus teaches that the deeper meaning of the commandment not to kill is to not even be angry. The commandment against adultery is similarly broadened to proscribe even lustful looks. The commandment against swearing false oaths is superseded by warning against swearing oaths altogether. The just retaliation of "an eye for an eye and a tooth for a tooth" is replaced by no retaliation whatsoever. Clearly these teachings of Jesus change the focus of righteousness from an external standard to an internal standard. It was no longer acceptable to refrain just from committing a bad act, the refraining was expanded to include bad intention and attitude as well.[33]

Another example of the revolutionary and trans-humanity heart of Jesus' teaching can be found in the juxtaposition of the Great Commandments and the Parable of the Good Samaritan. Luke 10: 25-28 is the familiar dialogue between the "scholar of the law" and Jesus. The scholar asks Jesus, "What must I do to inherit eternal life?" Jesus responds with a question, "What is written in the law?" The scholar replies, "You shall love the Lord, your God, with all your heart, with all your being, with all your strength and with all your mind, and your neighbor as yourself." Jesus assents, "You have answered correctly; do this and you shall live." Although sometimes in homilies this gem of teaching is enough to polish, the conversation goes on. "But because he wished to justify himself, he said to Jesus, 'And who is my neighbor?'" Jesus replies with the Parable of the Good Samaritan. In this parable, a man is robbed and beaten on the road from Jerusalem to Jericho. Both a priest and Levite pass him by without helping, possibly motivated by a desire to maintain ritual cleanliness. But a Samaritan traveler did stop and help both physically and financially. In this parable it is important to note that the person who gave aid was not an emotionally neutral character. Samaritans were looked down upon by the Jewish residents of Judah. At the end of the story, Jesus asks the scholar which passer-by was the neighbor, allowing the scholar to choose among those within his tribe who ignored the victim, or the out-of-tribe stranger who showed kindness. The scholar rightly answered, "The one who treated him with mercy." Rather than just telling the scholar that he had chosen correctly, Jesus' closing remark is a call to action: "Go and do likewise." In this exchange, Jesus shows that the concept of neighbor extends beyond borders or tribe, and that moral action is more important than ritual purity.

It would appear that Jesus intended his lessons, and the way of life he taught, to be concrete rather than abstract. This lesson is hammered home in Matthew 25: 31-46, sometimes referred to as "The Judgment of the Nations."[34] From the throne of judgment, Jesus tells those who are selected for eternal life:

> Come, you who are blessed by my Father. Inherit the kingdom that was prepared for you from the foundation of the world. For I was hungry and you gave me food, I was thirsty and you gave me drink, a stranger and you welcomed me, naked and you clothed me, ill and you cared for me, in prison and you visited me. Then the righteous will answer him and say, 'Lord, when did we see you hungry and feed you, or thirsty and gave you drink? When did we see you a stranger and welcome you, or naked and clothe you? When did we see you ill, or in prison, and visit you?' And the king will say to them in reply, 'Amen, I say to you, whatever you did for one of these least brothers of mine, you did for me.'

Jesus' moral teachings were profound in the first century and are still profound today. In fact, in the broad view, Jesus' life on earth was neither very long nor very long ago. On our toothpick path his 33 year life, just 33 toothpicks, is less than three inches long. Gazing back at our 18,000 miles of path, we only have to look back 14 feet to see the time of Jesus. The course of history tells us that Jesus' lessons have not yet been completely learned. But there have been other voices that have repeated, clarified and contextualized Jesus' radical demands, demands to strive for his own example: humanity infused with divinity.

Chapter 6

Development, Dissemination and Fragmentation

Go, therefore, and make disciples of all nations, baptizing them in the name of the Father, the Son, and the Holy Spirit, teaching them to observe all that I have commanded you. And behold, I am with you always, until the end of the age.

Matthew 28: 19, 20

Following the Axial Age and including the time of the life of Jesus, a phase referred to here as development, dissemination and fragmentation began. In this phase the religions which came into existence during the Axial Age were further developed, that is, following generations sought to better define the principles and practices that characterized the movements. This broadly includes what we would call both dogma and ritual. At the same time the movements were disseminated; they were spread both within and outside of their originating cultures. Redefining the religions in new cultural terms and new social situations resulted in continuing developments in the movements. Expansion of the boundaries of technology and geography contributed to these changes. The developments were far from homogeneous. Within each tradition there were currents of change and accommodation to new situations and counter-currents of holding fast to particularities of earlier versions of the religions.

These counter-currents as well as differing understandings of the developing bodies of dogma ultimately resulted in conflict and fragmentation. Although a human tendency might be to search for black and white in these countless internal conflicts, there were undoubtedly people involved on all sides and at all junctures who were conscientiously certain that right was on their side. It is fair to say that the impulse to develop, disseminate and fragment began shortly after each of the respective movements began.

Although the seeds of the world religions sprouted in the Axial Age, two important religions began branching away from their roots soon after. The first was Christianity and the second was Islam; both branches from the tree of Hebrew tradition. In this chapter we will briefly examine the course of these and highlights of other movements in the period.

An interesting aspect of this period is that within each of the faith traditions there was some impulse to look inside. Each tradition had their own mystics, those who through a variety of meditative practices sought a transcendent connection with something outside of themselves by looking within themselves. Common fruit from these various strains of mysticism will also be examined in this chapter.

Christianity

St. Paul and Jesus' apostles were the first movers in the remarkable spread of Christianity. The 27 books of the New Testament were likely written in the period of about

50 to 150 C.E., contemporaneously with the early spreading of the Gospel throughout the Roman Empire and the then-known world. Christian persecution within the Roman Empire continued until Constantine's Edict of Milan in 315 when Christianity became a permissible religion. At the same time, development of Christian doctrine was underway, often precipitated by debates which helped to define orthodoxy. For example, the early Christological debates which explored the nature of Jesus' humanity and divinity sparked considerable controversy.[35] Extreme positions in these debates were ultimately considered heresy, but from a positive perspective, these debates helped to clarify Christian understanding of Jesus' identity and resulted in a common creed.

Although the spread of Christianity, unprecedented in human history, must be seen as a bright light in the story of humanity, fragmentation is a dark cloud. The first such major division was the East-West schism in 1054. The Protestant Reformation movement began with Luther's nailing of his 95 Theses to the door of the Castle Church in Wittenberg, Germany in 1517.[36] This tendency to divide has continued; in fact it has dramatically accelerated. Although there is no single common factor which has caused fragmentation, disagreement on dogma and practices of the leadership deemed to be improper seem to be two important factors. When exacerbated by interests of power, politics and personal agendas, fragmentation has occurred. It is estimated that at least 38,000 Christian denominations with their own constitutions now exist, not counting single-church entities.

Despite the very human tendency to disagree and fragment, Christianity as a whole has been the largest religious movement in human history. The broad establishment of schools, hospitals and relief organizations make Christianity one of the most significant forces ever in the development of humanity and the alleviation of human suffering.

Islam

Of the major world religions today, Islam was the most recent to generate and develop. It is included in this chapter since it is a post-Axial Age development of Judeo-Christian roots. Muhammad was born in 570 in Mecca which is located in present day Saudi Arabia. According to tradition, he began receiving revelations at age 40. He is considered in his religion the last great law-bearer in the series of prophets which includes Abraham, Moses and Jesus. Although not an axial figure, the socio-political situation of his time and place was akin to Axial Age motivating forces. Conflict was violent and frequent among Arab tribes and Muhammad's movement sought to unify the tribes under one system of common beliefs. Muhammad adopted and adapted common roots with the Jewish people tracing his Arab origin to Abraham's son Ishmael, brother of Israel's father, Isaac. Muhammad was also familiar with Christian teaching and adopted and adapted select parts of the Gospels.[37]

Regarding the Jewish and Christian scriptures, Muhammad posited that at inception they were from God, but that through the centuries they had been corrupted through error in translation and editing. In contrast, he maintained that the Qu'ran was dictated to him directly by the angel Gabriel word-for-word in Arabic, never to be corrupted by translation into another language.

The name of the religion founded by Muhammad, Islam, means "to submit," understood to mean submission to God's will. Such submission was expected to lead to harmony among tribes, love among family members, and compassion for strangers. Violence was generally considered only as a last resort in protection from violent oppression perpetrated by outsiders.

In terms of human rights, it is important to note that Muhammad's intentions were not only abstract and religious, but concrete and civil. His Constitution of Medina in 622 was one of the earliest examples of human rights as an integral part of civil law.

Like other religions, fragmentation has been a factor in the Islam tradition. The first and principal division occurred immediately upon Muhammad's death. This division resulted in Sunni and Shiite factions which still exist. Today, there are about 940 million Sunni Muslims, generally considered to be more liberal and 120 million Shiite, generally considered to be more conservative.[38] Turkey and Saudi Arabia are examples of predominantly Sunni countries and Iran is an example of a predominately Shiite nation. Beyond Shiite and Sunni there are other smaller divisions. Within the two major divisions there are further formal divisions, particularly among Shiites. Some formal divisions among Sunnis also exist. Importantly, with no central authority, local Sunni leaders have considerable latitude in interpretation and application of the teachings of their faith.

Another significant point to make about Islam is that it shares a key characteristic with Christianity differing from other world religions: the inherent compulsion to spread the faith. Both Muslims and Christians believe that their messages of salvation are intended for the whole of humanity. This motivation has provided impetus for the remarkable spread of both. It has also been the source of conflicts: conflicts of ideas, conflicts of violence.

Mechanism of Fragmentation

Although there are a variety of factors that can contribute to fragmentation and there is no single process that can define how and why fragmentations occur, there are some denominators that are fairly common. Following is a sequence that portrays how some fragmentations occur.

1. A situation changes, new event occurs or a new threat is realized.
2. Two polar options for dealing with the problem are identified. One option might involve continuing to do the same thing as always or even returning to an older way of doing things. Another option might be to change to adapt to the new situation.
3. Leadership for both of the opposing ideas emerges and the group ideologically divides siding with one leader or the other.
4. The opposing ideologies are refined. Each group hones its view of the "rightness" of their position and the "wrongness" of the other.
5. Defining those of alternative opinion as the "other" homogenizes the group. Ideological dual silos inhibit and ultimately preclude constructive discourse between people of opposing opinions.
6. Expanding on the "wrongness" of the "other," the

"other" is vilified and demonized. That is, the view of the "other" intensifies from well meaning supporter of a legitimate alternate opinion to supporter of mistaken opinion, to supporter of an incorrect opinion, to supporter of an immoral opinion, to an immoral person, to a person against God and God's will.
7. The opposing parties decide that they can no longer live with each other, since each sees the other as a manifestation of evil. Leaders fan the fires of emotion in their followers.
8. The groups physically divide or continue to live together on divided terms. In some cases they seek to eradicate the other. Initial incidental violence can escalate to purposeful murder.

There are an infinite number of possible scenarios for how groups of humans can fragment; certainly not all involve violence and many have nothing to do with religion. As demonstrated above, fragmentation can occur without religious, political or ethnic identities, although these factors are often enlisted for support.[39]

Other Important Characteristics of the Period

In many ways it is impossible to concisely summarize the last two thousand years of human experience in simple terms. During this period, previously unknown areas of the world's surface have been explored, mapped and colonized. Colonization has generally been better for the colonizer than for those colonized and a wide variety of human atrocities were committed. Learning to harness resources and technology, the Industrial Revolution resulted in improvements in the standard of living for some, but often at the expense of others. The rise of technology since then has accelerated. Communication, travel and living conditions for the "haves" continue to improve, but still often at the expense of the "have nots." The importance of tribe has generally waned while the power of large industrial-based nation-states has generally waxed. Democracy has generally replaced monarchy as the predominate form of government in the West. Such forms of government have broadly provided increased levels of personal freedom, but not without struggle. Larger nations and advancements in technology have led to larger wars with higher human costs.

Several of these trends can be seen in the history of the United States. From a geographical standpoint, we are a product of the colonization efforts of Great Britain, France and Spain. Early settlement is often credited to the Pilgrims who left Britain on religious grounds, a fragmentation. The Revolutionary War was a fragmenting movement, carving our country away from the British Empire. Our Civil War was an unsuccessful effort to fragment on both ideological and economic grounds. Throughout the early history of our country there was an impulse to disseminate the faith as European colonists took their form of civilization and religion west. We have also played our part in development. This is obvious from an industrial and technological standpoint, but also true from the standpoint of theology. There have been many important U.S. theologians from a number of different Christian traditions.

Issues associated with the pros and cons of human advancement will be discussed more broadly in the next chapter, but first we will turn to discuss some important non-fragmentary threads of thought present among humanity in the same two thousand year period.

Mystics

> *Do you not know that you are God's temple and that God's spirit dwells in you? For God's temple is holy and you are that temple. 1 Cor. 3:16, 17b*

The portion of human history which has been largely described here as fragmentary also contains significant non-fragmentary thinking. As examples of these non-fragmentary impulses we turn to mystics. These are generally people who, within their given religious contexts, have devoted large portions of their lives to prayer, mostly meditative prayer. We will briefly survey general ideas of some Christian mystics, but a number of the insights are similar for mystics of other religious traditions. We will examine some ideas about God and/or creation which might be considered common denominators among them.[40]

Did you ever wonder what you could learn about yourself and God if you had the time or dedication to pursue this kind of life? The good news is that the mystical greats throughout history have hoped you would wonder and have left you their stories. This is true for Christian mystics and it is also true of mystics of other religious traditions. You might wonder whether there are any common denominators among their experiences: is there a common human "religious experience?" The answer is no. And yes. Although there are wide variations in experience of the mystics influenced by religious presuppositions, it is fair to say that there are, indeed, some common themes in their experience.

Although there is no commonly accepted list of mystical truths, the following are statements that are generally in agreement with the writings of the spiritual greats across religious boundaries:

1. All of creation is somehow connected.
2. This connection can be experienced internally as well as externally.
3. In Christian, Jewish and Muslim terms, this connection is with and through God. In Buddhist terms, the connection has no personalized name.
4. Everyday life is infused with holiness. God/connection can be experienced in everything. This includes not only prayer and meditation but art, nature, science, mathematics, cooking, walking, conversing, chance meetings, doing chores; everything noun and verb.
5. God is paradoxically unchanging but dynamic; creation reflects God's dynamism.
6. Our bodies are, indeed, the temple of the Holy Spirit.
7. Life in all forms is precious.
8. Life should be lived in harmony; with God, people, all of life, the earth, the cosmos.
9. Connection with Creation dials-down self interest and dials-up compassion.
10. Social justice is imperative: hungry people should be fed, homeless people should be clothed and housed.
11. Peace is attainable and divine and destined.

12. We are not God, but we are infused with God and God is infused with us.

The Chapter's End?

So where in time does this chapter end? There are problems associated with defining periods of history at all, and picking out starting and ending dates is particularly tricky. Considering the Axial Age, there were rare axial-type occurrences in pre-axial time. For example, the Code of Hammurabi was promulgated by the king of the same name in about 1760 B.C.E. It contained 282 statutes, each a sentence or two, many of which had bearing on the rights of the citizens of Babylon. On the other end of the Axial Age, as noted earlier, both Jesus and Muhammad fall past the end of Jung's definition, but both harmonize well with Axial Age themes. In regard to the development, dissemination and fragmentation period, we can summarily say that it begins at the close of the Axial Age, but where does fragmentation end?

Like looking at an elephant with a magnifying glass, the closer one gets, the poorer the understanding of the large animal as a whole. We must stand back as far as we can and determine whether the details that we are able to see reveal aspects of the big picture that we cannot yet bring into focus. In the next chapter we will examine some details of the past century that might shed light on where we are today. We will move our focus from the last 13 feet of the toothpick path to the last eight inches, now on the alter of St. Patrick's.

Chapter 7

Assessing the Trajectory of Humanity

> *I give you a new commandment: love one another. As I have loved you, so you also should love one another. This is how all will know that you are my disciples, if you have love for one another.* John 13: 34,35

Returning briefly to our path of toothpicks, recall the 18,000 mile path for the cosmos, while modern humanity has been around only about a quarter of a mile. Refining our scope even more, recorded history is represented by 22 feet of toothpicks and only 14 feet since Jesus' earthly ministry. We have noted that in the last 14 feet or so of toothpicks, since Axial Age movements began, the major faith traditions have experienced fragmentation, often dividing on ideological lines. The question that this chapter addresses is whether at this moment in history we are still in the same phase. There is a case to be made that humanity is at another turning point, another kind of Axial Age, more ready and willing to embrace the spirit of first Axial Age teachings.

To get started in this analysis we must ask ourselves, how have we, how has humanity, been doing in applying the Axial Age lessons? There can be little argument that technology has advanced. We may remember milestones of the history of technology from high school: the printing press in 1500's, the industrial revolution of the 1700's, the light bulb, telegraph, automobile and airplane in the early 1900's. In the last one hundred years the advent of space travel, world-wide communication, computers, and the internet would have boggled the mind of our Axial Age predecessors. The trajectory of technical advancement is positive and more revolutionary innovations are anticipated, and in fact, underway. Yes, it is easy to see that technology has advanced and is advancing, particularly from our climate-controlled homes in, arguably, the most technologically advanced nation on the earth today.

But is there another aspect of humanity that must be considered in evaluating whether any progress has been achieved? Certainly the moral aspect of human behavior is an important litmus test for whether humanity is on any kind of trajectory, either positive or negative. If the message of the Axial Age was for humanity to begin thinking and acting in more humane ways, how well has the message been heeded?

Unfortunately, the answer is not obvious. There were wars then and wars now. There was hunger then and there is hunger now. There was poverty then and poverty now. There were human inequalities then and there are human inequalities now. Still, the many voices in humanity's history which have echoed axial themes have not gone completely unheeded. Many such voices have risen up above the din of ambiguity and humanity has not been totally deaf.

There have been an extraordinary number of these kinds of voices in the last one hundred years that have served to inspire such change. Among these are Gandhi,

Nelson Mandela, the Dali Lama, Desmond Tutu, and Blessed Theresa of Calcutta, just to name a few. Each of these had profound impacts on the societies in which they lived, each in their own ways acting and inspiring others to act in ways that made Axial Age moral lesson concrete in a new time and place. Each is recognized for their contributions to human society not only in their own countries but on the world stage.

In our time and our country one such recent voice was that of Dr. Martin Luther King, Jr. Ta-Nehisi Coates, a writer who focuses on African-American issues and their impact on society draws an important distinction between Dr. King and others who have advocated for their African-American constituencies. Coates makes the point that Dr. King's insights had two parts.[41] The first part of Dr. King's insight was, of course, that there exists goodness among his constituents worthy of respect and equality. But the second part, and perhaps the more surprising part, was Dr. King's belief in the latent potential for goodness in all humanity. This led him to his vision that the plight of African-Americans in their struggle for equality in the U.S. would be ultimately corrected by the sense of justice and fairness of the white majority. Although Dr. King's strategy meant that more suffering would need to be endured before the right arguments would be respected in their right time, his remarkable prophetic voice told his generation that the time would come. The news media highlighted echoes of Dr. King's prophetic voice during the 2008 presidential race and the election of Barak Obama, the nation's first African-American president.

Although there have been a number of wise and influential people who have left the world a better place in the last one hundred years, it is interesting to wonder whether this represents a trend, or if such wise figures just randomly appear throughout human history. If there are such trends in human history, to what, if any, point are they directed?

The Concept of Trajectory

The early twentieth century voice of Jesuit scholar Pierre Teilhard de Chardin is particularly provocative in considering the trajectory of humanity. Within a Catholic Christian context, his voice piously speculated about how the unfolding process of the universe can be mystically wed to Christianity. Having studied both paleontology and religion in the first half of the twentieth century, Teilhard de Chardin was uniquely well positioned to reconcile the new scientific understanding of the long unfolding process of the cosmos with a personal relationship with Christ. He believed that this reconciliation requires a universe-wide and trans-time understanding of Christ. In his view, the universe is from Christ in Creation, evolves hand-in-hand with Christ, ultimately on trajectory to unity with Christ. In this scenario, creation becomes more Christ-like as the endpoint of Christ-identity is approached. Obviously, this view is as much mystical as it is scientific. Teilhard's science does not marginalize religion, but rather elevates it to a higher level with Christ as the canvass on which God paints the beauty of creation. The concept that humanity is slowly developing conscience and consciousness, and that the development is with and into Christ was revolutionary. So revolutionary, in fact, that his most significant book, The Divine Milieu, written in the late 1920's, was not brought to light until after his death some 30 years later.⁴² Teilhard's voice is one of joy and optimism: that humanity is, indeed, progressing and progressing towards unity with cosmic Christ.

The Case for a Positive Trajectory

Although the forces of development, dissemination and fragmentation are still at work, are there sufficient positive new movements which might indicate humanity is entering into a new period? Where can we start looking to see if such new movements exist? We know that if such movements exist that they will resonate with Axial Age themes, but take them a step farther.

Let's consider human rights. We have generally observed to this point a slow development in the concept of human rights. There were early roots in the 1760 B.C.E. Code of Hammurabi. The Ten Commandments followed by at least several hundred years. Jesus' beatitudes were first taught in about C.E. 24. Muhammad's Constitution of Medina was written in 622. King John's Magna Charta acceding civil rights to nobles was granted in 1215. Although this list is not comprehensive, it is fair to say that through history the concern for human rights developed slowly and locally.

Then things changed. Beginning in the eighteenth century both the pace and the geographical extent of interest in human rights dramatically increased. England's 1689 Bill of Rights was at the front of a trend. The U.S. Declaration of Independence in 1776 famously stated that "all men are created equal" with the inherent rights of "life, liberty and the pursuit of happiness." France promulgated its "Declaration of Rights of Men and the Citizen" shortly after in 1789. The U.S. Bill of Rights was amended to the constitution in 1791 insuring the freedoms of religion, speech and assembly. The first of the Geneva Conventions protecting human rights even in times of war was agreed to in 1864, the same year as the Leiber Code which initiated the International Red Cross. In 1919 the League of Nations established an international covenant which enumerated the human rights of life, liberty, freedom of expression, equality before the law along with other social, cultural and economic rights. Most significantly, in 1948 the United Nations established the Universal Declaration of Human Rights binding all nations of the world to a standard of conduct in human rights to which all agreed to be held accountable.[43] This Declaration positively asserts that all humans are free and equal, that they have rights to life, liberty, security of person, judgment under law, movement, family and property. It includes as basic rights the freedom of thought, exercise of conscience, religion and the right to work. It further asserts that slavery, torture and any form of discrimination are unacceptable in the human community. This is clearly the most significant human rights document in the history of the world.

Words are one thing and behaviors may be another so we should examine the track record. Let's think about the human right of freedom and its contrary, slavery. In the Axial Age slavery was a common practice. Soldiers of defeated armies routinely became slaves and entire populations were displaced to serve their victors. The Hebrew Babylonian captivity is a non-unique example. In New Testament times the buying and selling of people into slavery was a common occurrence, and those who study the Bible are surprised to find few teachings aimed to stop the practice. But, over the last two thousand years, most of humanity has come to both conclusion and practice that human slavery is not right and should be ceased. As we know from our country's history, this was not achieved without loss of human life and diminution of the country's economy. Has slavery been eliminated from the face of the earth? Not totally, but it is institutionally prohibited in all but a few nations, considered to be backward as well as immoral and inhumane.

What about the rights of children? Sons and daughters were considered to be personal property in the axial world. Under Roman law, it was not even a crime to murder one's children for any reason whatsoever. Certainly we have some improvement to discuss, but it has not come easily. Sometimes we humans do not respond to offense until the offense becomes so glaring that most any human with conscience can no longer silently tolerate it. Excesses in child labor during the early days of the industrial revolution fit this model. As the excesses of the practices of working children in slave-like fashion for endless hours under dreadful conditions came to light, laws were gradually enacted to protect them. Children's rights continue to be enhanced with safety, education and medical care accepted as basic rights in most developed nations.

A similar argument can be constructed for the advancement of women's rights, although such an argument made even one hundred years ago in the U.S. would have lacked teeth. Women's rights in the home, workplace and nation have seen notable advancements in the last century. In most developed nations there is no job or honor that a woman may not hold. In the U.S., the widespread sense of plausibility of Hillary Clinton's nomination and possible election to the presidency in 2008 raised the bar of women's equality to a new level. Throughout the developed world people are shocked by media images of women denied education in Taliban-influenced areas of Afghanistan. Just the fact that we are shocked speaks volumes about our notions of what humanity is about.

In that the ultimate victor of the most recent U.S. presidential race was a person of African-American descent, another inequality, ethnic bias, was positively impacted. Although in the U.S. we often cast this problem in a white-black or white-Hispanic frame, on a planet-wide basis the problem is much broader and might be another example of left-over tendencies towards tribalism. For example, ethnic differences have undoubtedly played a large part in the tensions in the Middle East. Arab-Jewish and Arab-Persian ethnic differences contribute to the low flashpoint of violence in the region. Violence in Bosnia and the region as a whole in the 1990's had significant contributing ethnic factors.

In the world of human ideas, one idea that is worthy of note in this optimistic section is the concept of the common good. The concept certainly had Axial Age roots, but the bloom came much later. Although it is difficult to nail down exactly where and when this concept was first heard spoken from the stage of humanity, it had a most cogent voice in Thomas Aquinas in the 13th century.[44] Following Aristotle, Aquinas noted that people's activities are directed to good ends, although some good ends are better than others. Practical examples of this are obvious enough. If one is hungry, eating a piece of pizza could be considered good. It can be both nourishing and taste good. Eating a second piece might also be good. Eating a third and a fourth piece might be considered less good, possibly adding to an unhealthy weight problem and even keeping someone who is truly hungry from eating. Having a warm coat in winter is good, but if you have five and others have none, the extra coats are not achieving their highest good in your closet. Catholic moral teaching goes so far as to say that the extra coats don't even really belong to you, so give them away. Aquinas extrapolates his argument for the common good to the greatest good: that every human activity should be directed towards achieving the greatest good possible considering the good of the community as well as the individual.

Although Aquinas' eloquence about the common good is rarely heard in common conversation, its practice seems to be more and more widespread. There are hundreds of organizations which encourage sharing on scales ranging from world-wide to local neighborhood. Called NGO's for non governmental organizations, some of these world wide agencies like the Red Cross-Red Crescent and Catholic Relief Services are often the first responders to instances of natural or political disaster. There are dozens of other such agencies which address human suffering in its many forms: hunger, poverty, disease, homelessness, political displacement and oppression. Amnesty International, Doctors Without Borders, Heifer International, Bread for the World and the personal initiatives of former presidents Carter and Clinton are just a few examples of applying the principles of the common good on a world scale. It is fair to say that there have never been as many such organizations and initiatives in the world as there are today.

Focusing on disease, consider the huge investments of both money and promising scientific and medical careers dedicated to eradicating causes of human suffering and death. The American Cancer Society and The American Heart Association are examples of organizations which have already contributed to help large numbers of people and show reasonable hope of significantly improving the quality and length of human lives. There are dozens of other similar organizations which hold genuine promise of making significant strides in reducing the suffering of those afflicted with their targeted diseases.

The notion of freedom is another concept that resonates with us as a basic human right. The founding fathers of the United States considered "life, liberty and the pursuit of happiness" rights that were and are "unalienable" from being human. The concept of political freedom is one that would have met with Aquinas' approval as well, and has met with approval among Christian leadership. It is important to note, though, that our form of freedom through a democratic form of government is neither wholly American nor wholly Christian. Pioneered by the Greeks and employed sporadically in the Roman Empire, the democratic form of government burns brightly, but not exclusively in the U.S. The largest democracy on the planet is in India, predominantly Hindu. The U.S. is the second largest and the third is Indonesia, predominantly Muslim. The democratic form of government that we enjoy is not flawless, by any means, but it seems to be the best system humanity has devised yet to allow for human goodness to emerge.

In the last 100 years, significant human attention has also turned to protecting the resources of the earth. Trying to rein-in the damage of unfettered industrial development, humanity has realized that we can spoil our own nests in a variety of ugly ways. We have come to recognize that we can spoil water, plant life, animal life, natures' beauty, even the air that we breathe. Not only have we come to realize that some of our actions have resulted in damage to our environment, we have taken some action to reverse the course of some ecological problems. In the U.S., for example, clean air and clean water legislation have resulted in significant improvements. Measures to protect endangered species have been enacted and have, in many cases, succeeded in altering the course of demise of some populations. The Bald Eagle, for example, once threatened with extinction due to the human enterprises of expansion and hunting are now much more commonly seen and their populations are expanding. There are many different examples of the emerging concern of humans for the environment. We are concerned about the loss of wetlands. We are concerned about the loss of rain forested areas. We are concerned about global climate change.

We can see that there is a long and significant list of issues that we humans are fighting against including hunger, poverty, disease, political and personal oppression, corruption, violence, despoiling the environment, even in-humane treatment of animals. In addition to these things which we humans actively fight against, there are many good things we fight for. Concerning fellow humans, there is an emerging sense that every human deserves to be treated by all others with profound dignity and this has given rise to fights for education, health and economic prosperity. There are many more things on the list of things that humans fight for including the various arts, parks, recreation, museums and celebration of heritage. We fight against social injustice and we fight for the dignity and joy and beauty that we see as part of our humanity.

Theologian and philosopher Rene Girard joins in the position that moral advancement of humanity seems obvious:

> Our society abolished slavery as well as serfdom. Later has come the protection of children, women, the aged, foreigners from abroad, and foreigners within. There is also a battle against poverty and 'underdevelopment.' More recently we have made medical care and the protection of the handicapped universal. Every day we cross new thresholds. When a catastrophe occurs at some spot on the globe, the nations that are well off feel obliged to send aid or participate in rescue operations. You may say that these gestures are more symbolic than real and reflect a concern for prestige. No doubt, but what era before ours and under what skies has international mutual aid constituted a source of prestige for nations?[45]

In summary of this view, humanity seems to be on a generally positive trajectory. Issues in human rights that weren't even seen as problematic two thousand years ago have been recognized and are being addressed. Although the practice of war continues, there is a wide recognition that peace is better and war should be a last resort and only applied in limited circumstance, generally to protect the loss of innocent life. The notion of the common good is so infused into our society that most children in developed countries grow up knowing that attending to the needs of others is the right thing to do.

We have established that humanity is young; we humans have been part of creation for just a cosmological moment. In the view of a positive trajectory of humanity, we might consider the metaphor of a growing child. As the child grows and learns, at some point the child has enough control to become the agent of their own destruction. For parents who have raised children through the teenage years, handing over the keys to the car for their first solo driving expeditions can be anxiety producing, if not horrifying. We know that most such trips in the family car with a teenager behind the wheel end without incident. But we also know full well that a few end in tragedy. Are we, humanity, in our teenage years? Are we walking up to the edge of the cliffs of nuclear and biological extinction and peering over with a little too much bravado? Will we be smart enough to let our developing good judgment outweigh our childish selfishness and false sense of invincibility? Can we survive our fitful teenage years with all of its emotion and drama and risk without harming ourselves and creation around us? Will we, as humanity, grow into responsible adulthood or are we destined in every generation to live out the danger of our teenage impulses?

A Significant Note of Caution

Despite the large number of human achievements on moral as well as technological fronts, there is a rational view that is less optimistic. This view posits that there is still so much injustice and large-scale human unethical behavior that it is inappropriate to view the positives alone and use them to chart any trajectory whatsoever. This position highlights the reality of some modern heinous human acts which are of a larger scale than anything that has come before in human history. Among these are genocide, unleashing weapons of mass destruction on civilian populations and institutionalized killing of the unwanted. Such examples are Nazi mass murders of Jews during World War II, U.S. use of nuclear warfare in World War II, genocide in Africa, and laws allowing for abortion and euthanasia. Such a view is not necessarily pessimistic, but assesses human moral behavior as ambiguous.

There is certainly a large measure of practical truth in the position of human behavior being morally ambiguous. We may talk about much good on a wide range of issues, but if the proof of the pudding is in the tasting, in many cases the pudding does not taste good. For almost any human rights issue one can name, there are still significant problems. Most societies decry slavery, but it still exists in its original forms in some places in the world, and it has developed into more subtle forms in others. Sex slavery, with women from less developed countries sold into lives of prostitution is an example of a form of slavery which, tragically, is alive and thriving in the world. Or consider worker's rights. Although workers rights, including women's and children's rights are espoused from many corners of the globe, globalization has resulted in jobs often flowing to places where rights are ignored and the cost of labor is the lowest.

Another example of moral ambiguity is when new rights for one group diminish rights for others. One such contemporary tragic example has to do with the rights of women. The movement which has rightly helped to advance issues of equality for women has also served to advance the issue of abortion. In achieving the right to control their own bodies, women have achieved the associated right to end the lives of their unborn children. The number of abortions performed annually in the U.S. alone would have outraged not just previous generations of Christians, but adherents to every other major world religion. This is a most serious contemporary dilemma, perhaps the most serious moral dilemma of our time. Certainly it is a moral good for every human to have responsible control of their own bodies. Certainly allowing unborn children to thrive in their humanity and enjoy the blessing of life, liberty and the pursuit of happiness is also a moral good. This sad collision of conflicting goods has yet to be reconciled by humanity. Human response to this dilemma is, at best, ambiguous.

Still another example of ambiguity is in efforts to solve the problem of world hunger. As mentioned previously, there have been significant efforts to curb world hunger, and there is evidence that the situation is modestly improving due to this selfless outpouring of help from a very large number of caring people. Clearly this is an example of moral good at work. But there is another side, a much less positive side, to the story. Experts in world food supply and the problem of hunger have come to an astonishing conclusion: there is already enough food growing capacity for all of the hungry in the world to have at least minimal nourishment.[46] No one on earth needs to starve to death. Further, if some small additional steps were taken to enhance the food producing system, everyone could be well fed. Why isn't this happening? The system is complex and there are a variety of contributing factors.

S. Craig George

A significant amount of acreage is non-productive due to political upheavals, payment for non-production, and lack of labor supply in some third world countries. The aids epidemic in Africa, for example, has virtually eliminated segments of the farming population. Patent rights in seed production limit poor countries from obtaining the best yielding seeds. Corrupt individuals in powerful positions sometimes reroute donated food supplies for their personal gain. These are just a few of the many contributing reasons that hunger remains a problem on the stage of humanity. Our steps forward to date are surely morally good, but the fact that we are so far from achieving what is clearly achievable is, again, ambiguous.

One important thinker who is skeptical of too positive a conclusion about the current direction of humanity is world class scholar and theologian Carl Ratzinger, Pope Benedict XVI. Influenced by the horrors of World War II, Benedict considers such atrocities as mass murders in German concentration camps and the unleashing of nuclear weapons so terrible and on such a large scale that any perception of a trajectory of developing human goodness must be called into serious question. The Pope's thoughtful critique of the optimistic position also takes into account new negative developments. Among these are right to life issues including legitimization of abortion, euthanasia of the elderly, and embryonic stem cell research. Another major area of the Pope's concern has to do with the dangerous implications of modern technology. The human race has the capability to self-destroy by nuclear, biological and chemical means.[47]

A reasonable question to be asked about the many inhumane examples of modern human behavior is whether the in-humane in the human is destined to be with us always. Are humans so influenced by both inherited and developed selfishness and disregard for the "other" that just a few people acting on such urges will always pose a threat to human exploitation and even self-destruction of humanity?

There are several negative human tendencies which merit at least brief discussion. The first is that of scapegoating. Philosopher and theologian Rene Girard describes the "single victim mechanism" as a primary human social mechanism for coping with a problem situation. According to his view, in a problem situation a group tends to gravitate towards singling out a person to blame for the problem at hand. The person is murdered or banished which somehow gives the group a sense of having self-cleansed, at least until the next adverse situation arises. Girard contends that this is a, if not the, primary manifestation of evil among humans.[48]

A second deconstructive human tendency is to distinguish between "one of us" and "the others." Possibly left over from tribal impulses, this gives rise to ethnic-based conflict, religious-based conflict and operates on a variety of social levels. Identifying another person as different is not the problem, but rather the conscious or subconscious devaluing or demonizing of the "other." In looking at other countries, people of different ethnicities, different religions, different political affiliations or even different levels of education, it is usually easy to point out perceived flaws in the "other" to prove to insiders that the outsiders are "bad." We see this all-too-human tendency played out over and over. We are shocked and horrified when life threatening spectator violence breaks out at sporting events between supporters of rival teams.[49] Shakespeare and Rogers and Hammerstein have portrayed how this tendency can play out in needless tragedy in Romeo and Juliet and its adaptation in West Side Story.

Hoarding of resources is a third significant deconstructive tendency. With both individual and collective manifestations, those who have the power to do so tend to amass more than they need for comfortable survival. Possibly as a leftover from our survival instinct to hoard in a time of plenty for a time of famine, this instinct can result in stark contrast between the haves and the have-nots. This instinct coupled with the power that control of resources can bring tends to inspire the haves to fight for status quo in suppression of the have-nots. We have noted in earlier chapters that preserving the status quo is not a law of the cosmos, our planet or the nature of life. Still, powerful humans and human institutions guard their power with their lives, and, unfortunately, the lives of the less fortunate that they control. A human tendency is that our grasp is so tight on the status quo that we cannot release it to take hold of something much better.

The Second Axial Age

Weighing the positives against the negatives of worldwide human behavior is difficult, at best. We are surrounded by moral ambiguity, but the voices of optimism are gaining in confidence and number. Noting the positive signs of our times the idea that humanity is entering its second Axial Age has surfaced. According to this view, the positive developments, particularly in the last one hundred years or so are so profound as to suggest that humanity is at another significant turning-point. Karen

Armstrong, an important contemporary voice on world religions and historical trends, suggests that the roots of a second Axial Age may have first sprouted into view even as long as three hundred years ago.[50] The flowering of the concept of freedoms within society seems to weigh heavy in her analysis. Certainly such ideas gave rise to individual freedoms under democracies that were generally absent under the system of monarchies and feudalism. In Armstrong's view, these broad national movements paved the way for forward-thinking states to provide morally-based rights to their citizens.

In such a view, the snowball of enacting basic human rights for all has reached such a proportion that it will not stop and cannot be ignored. Although support for the ideas of a Second Axial Age could only be viewed as incomplete at this time, the concept may continue to gain support as further strides are made to insure the basic rights for all people on all parts of the globe. Unfortunately, such movements are seen most clearly in the rear view mirror, so future generations will be in a better position to judge whether humanity today has turned to face a better direction.

Projecting A Path of Hope

Unfortunately, the ambiguities of the human situation prevent us from plotting humanity's future trajectory with laser precision. Still, it might be important for each one of us to hazard our own guess. Our guess of trajectory could and should influence the choices we make in our lives each day.

Here is the outline of the path of humanity that appeals to this author. Rational human beings have been around a very, very short time in the 13.7 billion year unfolding of creation. The idea that this unfolding wonder of a cosmos with the emergence of life has been a totally random process devoid of any trajectory or spiritual underpinnings lacks convincing logic, spiritual experience, and majority opinion. Along the way of unfolding, a sense of morality, a sense of what it means to be a proper human seems to have developed among us. From a Christian viewpoint, these moral ideals were taken to a new level by Jesus. These teachings and enduring Spirit have caused followers to be the largest group of humans ever to exist on the face of this planet. The second largest group does not totally disagree with much of our moral teaching. There is significant evidence that humanity is, indeed, becoming more humane although there is still a long way to go and some human lapses is moral judgment are horrific. We may, indeed, be on the leading edge of a Second Axial Age in which humanity will take another significant step in achieving our potential and destiny.

Along this path, more and more of us will seek and find God, in churches, yes, but the foundation will be discovered inside our own beings. We will not find it difficult to access this divine mystery, but will find it increasingly difficult *not* to see Divinity anywhere we look. We will increase in our sense of interconnectedness with all of creation. We will live accordingly.

Humanity can come to see a bigger picture: that we may have eons of time ahead of us to achieve our potential. The eons of creation behind us are vast, why not the eons ahead, too? It may take us a long, long time to fully practice what Jesus preached, and that may be the plan.

We may come to fully internalize the concept that

every moment and every life is precious, each contributing an essential fragment of thread in the breathtaking beauty of the tapestry of creation.

In the end we will come to agree with Teilhard and the mystics. We will see that we are from Christ, we are always one with Christ, and our destiny will be fulfillment in Christ. This is the ending our story demands.

Chapter 8

Epilog: Playing Our Role in the Story

Is such the fast that I choose,
a day to humble oneself?
Is it to bow down the head like a bulrush,
and to lie in sackcloth and ashes?
Will you call this a fast,
a day acceptable to the Lord?

Is not this the fast that I choose: to loose the bonds of injustice,
to undo the thongs of the yoke,
To let the oppressed go free,
and to break every yoke?
Is it not to share your bread with the hungry,
and bring the homeless poor into your house;
When you see the naked, to cover them,
and not to hide yourself from your own kin?

Then your light shall break forth like the dawn,
and your healing shall spring up quickly;
Your vindicator shall go before you,
the glory of the Lord shall be your rear guard.
Then you shall call, and the Lord will answer;
you shall cry for help, and he will say, Here I am.

Isaiah 58: 5-9a

You who believe, be God's helpers. As Jesus, son of Mary, said to his disciples, 'Who will come with me to help God?' The disciples said, 'We shall be God's helpers.' The Qur'an: 61: 14a51

S. Craig George

The question of humanity's trajectory is interesting, but the following question is more important: Are we living in a way that respects the truth and that can lead the world to be a more Christ-like place? Are we playing our parts well in the unfolding drama of God's creation? If we're not sure, can we imagine what playing that part would be like?

Let's think for a minute about response to tragedy. In particular, let's think about response to the tragedy of September 11, 2001. You've undoubtedly heard hours upon hours of commentary about the attack, the heroism involved in the initial response, and military response thereafter. But you may not have heard about some other important responses. In her book, *A New Religious America*[52], Diana Eck describes some aspects of response that deserve widespread recognition and serious consideration. One such aspect was that there was a broad and timely Muslim condemnation of the attack that went largely unnoticed. Before the sun had set on that day that we will never forget, eleven important Muslim organizations had issued a joint statement condemning "viscous, cowardly acts of terrorism" and calling for the "swift apprehension and punishment of the perpetrators." In subsequent days, a sea of individual Muslim scholars and leaders raised their voices to condemn the acts that they believed had nothing at all to do with the teachings of their faith. Our government, our press and our population in general did not seem very interested in this response; it received little attention.

Violent acts were played out against Muslims, yet there were others among us who said, "No!" In Eck's book, she reports that such violence was threatened against a mosque in Toledo. A local Christian radio station became aware of the threat. They denounced the intention of violence and encouraged listeners to assemble outside the mosque at the normal time of worship to pray for peace. Two thousand people, mostly Christians, responded. They came together and held hands in a ring around the mosque and prayed for protection while their Muslim neighbors prayed inside to Allah. This is the kind of Christian, the kind of human, we all can be. This is the kind of human we all must be.

In studying the roots of the tragedy of September 11 and other violent conflicts, we can see more clearly the conditions which lead to terrorism. Factors which contribute to such behavior include economic disparity, lack of hope for improving the status quo, the tendency to demonize the "other," and the ability of extremists to separate some children from the world at large and convince them of the plausibility of violence and terrorism as the only route of achieving "their" God's aims on earth.

These conditions for hate and violence among all societies and religions can and must be cut-off. The Christian community in general and the Catholic Church in particular have positively recognized this opportunity for peace. In a 2009 article, "Blueprint for Peace," Archbishop Timothy Dolan highlighted the emphasis of Christian leadership in fighting poverty to achieve peace.[53] "The conditions in which a great number of people are living are an insult to their innate dignity and as a result are a threat to the authentic and harmonious progress of the world community." This quote of Pope John Paul II, echoed by Pope Benedict XVI in his 2009 New Year's message to commemorate the World Day of Peace, sounds the clarion call to stop violence at its source.[54] In this quote John Paul purposefully used the phrase "progress of the world community." Benedict goes on to show how the recipe for violence should be cut-off before it gets a chance to begin: "One of the most important ways of building peace is through a form of globalization directed towards the interests of the whole human family. In order to govern globalization, however, there needs to be a strong sense of global solidarity, between rich and poor nations, as well as within individual countries, including affluent ones."[55] As former head of Catholic Relief Services, Archbishop Dolan has seen the benefits of sharing resources and love where the rubber meets the road: "The hunger and poverty caused by the global food crisis are placing huge strains on the social ministry of the church, but the work of the Catholic Relief Services and other groups to bring relief to those most effected by the crisis is an example of how the church's message of peace can be brought to life."[56] This sense of an emerging human solidarity is not only gaining sway among religious thinkers, but entering the secular consciousness as well.

The global sense of solidarity which Pope Benedict and Archbishop Dolan encourage may be the linchpin for a positive future for humanity and our planet. This sense seems to be gaining momentum. Examples of growing solidarity include the world-wide concern for global climate change and the trans-national participation in relief efforts in response to natural events like the Haiti earthquake in 2010, hurricane Katrina and the tsunami of 2008. Even human induced tragedies like genocide in Darfur have received significant attention on the world stage.

Still, there are many issues where national self-interests seem to outweigh desire for peace. Currently concerns of nuclear armament of North Korea and Iran are among such potential flash-points. Palestinian-Israeli conflict is also a threat to world peace bearing all three of the dangerous factors of national, religious and ethnic differences. Peace in the sense of lack of violence is an extremely important issue, but just one of many peace related issues which will define the nature and future of humanity. Among other critical issues are the eradication of suffering from hunger, poverty, disease and social injustice. The respect for all life must undergird all human enterprise.

Given that many of these potential threats to humanity are of a global scale, as individuals we may feel that helping is beyond our reach. What can the individual do? The following items are ideas to help you engage in the advancement of creation; God's work in progress:

1. Love God, love all people.
2. Cultivate an interior life of prayer and meditation. Everyone can be, and should be, a mystic!
3. Seek to cultivate knowledge and goodness in yourself. Analyze your own thoughts and actions and cultivate virtue. Challenge yourself to grow in wisdom and skills which you can ultimately use to help others.
4. Become active in at least one area of social justice working to ease suffering of the poor, the hungry, the sick, the disenfranchised.
5. Be an advocate for peace. You can be an agent for peace wherever you are: your home, your job, your school, your community, your nation. Reach out

to understand and befriend the "other" until you understand that the "other" is not "other" at all.
6. Love God, love all people.

Is humanity on a positive trajectory or are we stuck forever in the angst and drama of a potentially self-destructive teen-like moral existence? The jury is out. No, more rightly, the jury is still in; the court is in session. Our lives are now the witnesses. We are all a mixed-bag of generally good intentions with some selfish or self-destructive behaviors. But in the main, is your life a witness to a positive trajectory or a witness to moral ambiguity? We know that our tendencies to serve self are hard-wired into us but we also have the amazing capacity to change. And we have Grace, the subtle urging from a source beyond us which quietly encourages us to play our part with compassion for our short time on the stage of the cosmos.

If we are aligned with the view of creation and humanity on a positive trajectory, we must realize that we are the agents of God in the advancement of God's program. We must be engaged. In this view we must revere our debt to the past and honor our obligation to the future.

Let's return to our toothpick path one final time. Lay out the number of toothpicks that represent your life. There are a certain number of toothpicks which represent your number of years so far, and a less certain number to represent your future years on the earth. If we're fortunate, the total will be about 7 inches or more, about 84 years more or less. Recall that we began our toothpick path on the altar of the Cathedral in New York. All of the toothpicks representing your life fit easily on the altar. Seven inches isn't much compared with 18,000 miles that came before us, but every life, every moment, is important. Every life can contribute, participating with God in moving the process of creation forward.

You do, in fact, live each moment of your life on the altar of the living God, on the stage of Creation. Each toothpick is precious. And here is our challenge: in each precious moment we may choose to work only for our own personal interest or the interest of humanity as well. Our opportunities to work with God in advancing humanity are both limited and sacred. We must use our precious time wisely. Our time on the path is short; our role in the story is crucial.

Animated and drawn together in his Spirit we press onwards on our journey towards the consummation of history which fully corresponds to the plan of his love: "to unite all things in him, things in heaven and things on earth."

Gaudium et Spes, Ephesians 1:10[57]

And all shall be well, and all shall be well, and all manner of things shall be well.

Dame Julian of Norwich

Notes, References and Further Reading

1. Biblical references are from the St. Joseph's Version of the New American Bible unless otherwise noted.

2. John Paul II, Encyclical Letter: *Fides et Ratio* (Washington, D.C.: United States Catholic Conference, 1998), n. 33.

3. Stephen W. Hawking, *A Brief History of Time* (New York: Bantam Books, 1988), 35-52. Although this book is somewhat dated, it is a classic in its concise explanation of complex concepts. For an explicitly Christian perspective see Denis Edwards, *Jesus and the Cosmos* (New York: Paulist Press, 1991) Also, see any contemporary physics text.

4. Stephen W. Hawking, *Black Holes and Baby Universes and Other Essays* (New York: Bantam Books, 1993), 115-123; Mario Livio, *The Accelerating Universe: Infinite Expansion, the Cosmological Constant, and the Beauty of the Cosmos* (New York: John Wiley, 2000).

5. Hawking, *Black Holes*, 119; Jurgen Moltman, *Science and Wisdom*, translated by Margaret Kohl (Minneapolis: Fortress Press, 2003), 164, 165.

6. Neil Shubin, *Your Inner Fish: A Journey Into the 3.5 Billion Year History of the Human Body* (New York: Random House, 2009); Ernst Mayr, *What Evolution Is* (New York: Basic Books, 2001) 12-73.

7. Karen Armstrong, *The Case For God* (New York: Knopf, 2009), 18-19, 29. In addition to Enuma Elish, the Akkadian myth "The Epic of Gilgamesh" from about 2000-1700B.C.E. seems to contain a remarkably similar account of the Noah flood story as well as parallels to Genesis-creation in which a serpent foils the first human's chance for immortality.

8. "Pontifical Biblical Commission Document on the Interpreta-

tion of the Bible in the Church" in *The Scripture Documents: An Anthology of Official Catholic Teachings*, translated and edited by Dean P. Bechard (Collegeville, Minnesota: The Liturgical Press, 2002), 244-315.

9   Ibid., 280.

10 Benedict XVI, *In the Beginning: A Catholic Understanding of Creation and the Fall*, translated by Boniface Ramsey (Grand Rapids, Michigan: William B. Eerdman's Publishing, 1995), 19-41.

11 Donald C. Johanson, *Lucy's Legacy: The Quest for Human Origins* (New York: Harmony Books, 2009).

12 Ibid.

13 Ibid., 253-268.

14 Spencer Wells, *The Journey of Man: A Genetic Odyssey* (New York: Random House, 2003) 182-183.

15 Michael Cook, *A Brief History of the Human Race* (New York: W.W. Norton, 2003); Cyril Aydon, *A Brief History of Mankind: 150,000 Years of Human History* (Philadelphia: Running Press, 2009).

16 Cook, *A Brief History*, 44-47.

17 Michael F. Roizen and Mehmet C. Oz, *You: Being Beautiful* (New York: Simon and Schuster, 2008) 133-142, 318-319.

18 Ibid.

19 Karen Armstrong, *A Short History of Myth* (Edinburgh: Canongate, 2005).

20 See *New American Bible*, introduction to the Pentateuch and introduction to Genesis.

21 John R. Sachs, *The Christian View of Humanity: Basic Christian Anthropology* (Collegeville, Minnesota: The Liturgical Press, 2001) 11-26.

22 Karen Armstrong, *The Great Transformation: The Beginning of our Religious Traditions* (New York: Knopf, 2006); Mark W. Muesse, *Religions of the Axial Age* (Chantilly, Virginia: The Teaching Company, 2007).

23 Joseph Blenkinsopp, *The Penteteuch: An Introduction to the First Five Books of the Bible* (New York: Doubleday, 1992) 1-30; John W. Miller, *Meet the Prophets: A Beginner's Guide to the Books of the Biblical Prophets* (New York: Paulist Press: 1987).

24 *The Principal Upanishads: The Essential Philosophical Foundation of Hinduism*, translated by Alan Jacobs (London: Watkins Publishing 2007), Kena Upanishad Part 4 n. 9, p. 27.

25 *The Dhammapada: Essential Teachings of the Buddha*, translated by Fredrich Max Muller (London: Watkins Publishing, 2006), n. 116.

26 Ibid., n. 169.

27 Ibid., n. 256, 257. See also Karen Armstrong, *Buddha* (New York: Penguin, 2001).

28 Mark W. Muesse, *Religions of the Axial Age*.

29 *Dao de Jing: A Complete Translation and Commentary*, translated by Hans-Georg Moeller (Chicago, Illinois: Open Court, 2007) n. 27.

30 Ibid., n. 31.

31 Elizabeth A. Johnson, *Consider Jesus: Waves of Renewal in Christology* (New York: Crossroad, 1990); Donald Senior, *Jesus: A Gospel Portrait* (New York: Paulist Press, 1992); James Alison, *Knowing Jesus* (Springfield, Illinois: Templegate, 1993).

32 Servais Pinckaers, *The Sources of Christian Ethics*, translated by Mary Thomas Noble (Washington, D.C.: The Catholic University of America Press, 1995) 134-167.

33 Ibid.; Timothy E. O'Connell, *Principles for a Catholic Morality* (New York: Harper Collins, 1990).

34 Ibid.

35 Richard A. Norris, *The Christological Controversy* (Philadelphia: Fortress Press, 1980).

36 Thomas Bokenkotter, *A Concise History of the Catholic Church* (New York: Doubleday, 1990); John Coakley and Andrea Sterk, *Readings in World Christian History* (Maryknoll, New York: Orbis, 2004).

37 Seyyed Hossein Nasr, *The Heart of Islam: Enduring Values for Humanity* (New York: Harper Collins, 2004): William Dudley, ed., *Islam: Opposing Viewpoints* (Farmington Hills, Michigan: Thomson Gale, 2004); Karen Armstrong, *Islam: A Short History* (London: Phoenix Press, 2005).

38 Estimates of demographics of religious groups are difficult, at best. An online resource, adherents.com, provides estimates of demographics of all world religious groups.

39 Amartya Sen, *Identity and Violence: The Illusion of Destiny* (London: W. W. Norton, 2007); R. Scott Appleby, *The Ambivalence of the Sacred: Religion, Violence and Reconciliation* (Lanham, Maryland: Rowman and Littlefield, 2000).

40 Harvey D. Egan, *An Anthology of Christian Mysticism* (Collegeville, Minnesota: Liturgical Press, 1996); Ursula King, *Christian Mystics: The Spiritual Heart of the Christian Tradition* (New York: Simon and Schuster, 1998); Kyriacos C. Markides, *Gifts of the Spirit: The Forgotten Path of Christian Spirituality* (New York: Doubleday, 2005). Also books by Thomas Merton and Thomas Keating are excellent contemporary presentations on the fruits of contemplation.

41 Ta-Nehisi Coates, NPR *Fresh Air*, Interview with Terry Gross 19 January, 2009.

42 Pierre Teilhard de Chardin, *Science and Christ*, translated by Rene Hague (New York: Harper and Row, 1968) 79-82; *The Divine Milieu* (New York: HarperCollins, 1960); *Christianity and Evolution*, translated by Rene Hague (New York: Harcourt, Brace and Jovanovich, 1971). From his work in paleontology, Teilhard concluded that life was in a constant state of change, generally to a higher order. He viewed creation in a constant state of "becoming," and moving from a material primacy to a spiritual primacy with an ultimate destiny of unity with Christ. Pp. 34, 79-82, 94. See also Beatrice Bruteau, *God's Ecstasy: The Creation of a Self-Creating World* (New York: Crossroad Publishing, 1997).

43 This brief yet profound document is available online at www.un.org/en/documents/udhr/.

44 Thomas Aquinas, *Summa Theologica* II-II q. 25, 6 ad. 2; q. 64, 2.

45 Rene Girard, *I See Satan Fall Like Lightening*, translated by James G. Williams (Maryknoll, New York: Orbis Books, 2001), 166.

46 Francis Moore Lappe, Joseph Collins and Peter Rosset, *World Hunger: Twelve Myths* (New York: Grove Press, 1998), 8.

47 Benedict XVI, "The End of Time," in *The End of Time? The Provocation of Talking About God*, translated and edited by J. Matthew Ashley (New York: Paulist Press, 2004), 9, 13.

48 Rene Girard, *I See Satan Fall Like Lightening*, 154-193.

49 Amartya Sen, *Identity and Violence*; R. Scott Appleby, *The Ambivalence of the Sacred*.

50 Karen Armstrong, *Suggestions for a Second Axial Age* (London: Polebridge Press, 2009, DVD) from Once and Future Faith Conference, Spring, 2001.

51 M. A. S. Abdel Haleem, *The Qur'an: A New Translation* (Oxford, England: Oxford University Press, 2004).

52 Diana L. Eck, *A New Religious America: How a "Christian Country" Has Become the World's Most Religiously Diverse Nation* (San Francisco: Harpers, 2002).

53 Archbishop Timothy Dolan, "Blueprint for Peace: Pope Benedict's Call to Fight Poverty" in *Commonweal*, 27 February 2009, Vol. CXXXVI Number 4, 12-13.

54 Ibid.

55 Ibid.

56 Ibid.

57 Pastoral Constitution on the Church in the Modern World, Gaudium et Spes, Vatican II, 7 December 1965, in Vatican Council II, vol. 1, The Concilar and Post Concilar Documents (Northport, New York: Costello, 2004), n. 45.

Index

A. afarensis 23
Allegories 18
Amos 38, 39
Aquinas, Thomas 46, 78, 79
Aristotle 46, 78
Armstrong, Karen 88, 89
Axial Age religions 36
Axial Age themes 38, 39, 47, 48
Axial Age timing 36, 48
Babylonian Exile 37, 38
Beatitudes 52-55
Benedict XVI 18, 86, 95, 96
Bible 15-18, 30-33
Big Bang 6, 8
Blueprint for Peace 95, 96
Buddhism 41
children's rights 77
Coates, Ta-Nehisi 73
Code of Hammurabi 69
Colonization 65, 66
Confucius 43
Constitution of Medina 63
cosmos, age 7
Dao, Daoism 42-44
de Chardin, Pierre Tielhard 74, 75, 91
DNA 23, 24, 27
Dolan, Archbishop Timothy 95, 96
earth, age 10
Eck, Diana 94, 95
Ecology 80, 81
Edict of Milan 61

Enuma Elish 16
Evolution 11
fragmentation mechanism 60, 64, 65
freedom 66, 76, 77, 80
Genesis creation account 5, 14, 21, 31
Genesis creation sequence 14, 15
God's Image 32, 33
Girard, Rene 82, 87
Great Commandments 56, 57
hard-wired human impulses 28, 29, 78
Hinduism 40
Hoarding 28, 88
Homo Sapiens 22, 24, 26
Homo Ergaster 23
Hubble, Edwin R. 6
human rights 75, 76
human trajectory 71, 72, 74, 82
hunger 85, 86
Isaiah 39, 93
Islam 62, 63
Ice Age 13
Jaspers, Karl 36
Judaism 37, 38
Judgment of Nations 57, 58
King, Jr., Dr. Martin Luther 73
Kongzi 43
Laozi 43, 44
literal sense of scripture 17
migration, human 24, 25
Muhammad 62, 63

Mystics 67, 68
Mythology 29, 30
Paleontology 11
Parable of the Good Samaritan 56, 57
Paul, St. 60, 61
Plato 45, 46
Protestant Reformation 61
Qu'ran 62, 93
Schism, East-West 61
Second Axial Age 88, 89
September 11, 2001 94
Sermon on the Mount 52, 54, 55
Sermon on the Plain 53
single victim mechanism 87, 88
slavery 76, 77, 84
Socrates 45
Summa Theologica 46, 78
timing of humans 22-24
timing of species 11, 12
Universal Declaration of Human Rights 76
Upanishads 40
Violence 29, 84, 87
women's rights 77, 78, 84, 85
written history 26
wu wei 44

About the Author

Craig George has degrees in both science and theology. With a first career in engineering and earth sciences and a second career in theology and writing, Mr. George is uniquely qualified to write about convergences between science and religion. He writes on a broad range of theological topics and directs adult education at his home parish. Additionally, Mr. George is active in a number of organizations and boards including the Academy of Mechanical and Aerospace Engineers at Missouri University of Science and Technology and the Catholic Library Association of Greater St. Louis.

LaVergne, TN USA
12 August 2010
192998LV00001B/9/P